General Chemistry Lab Manual

Updated Revised Third Edition

W. Lin Coker, III
Kimberly A. Elmore
Campbell University

Kendall Hunt
publishing company

All photos and line art illustrations throughout the text provided by W. Lin Coker, III and Kimberly A. Elmore

Cover image © Shutterstock.com

www.kendallhunt.com
Send all inquiries to:
4050 Westmark Drive
Dubuque, IA 52004-1840

Printed in the United States of America

CONTENTS

ACKNOWLEDGMENTS

This laboratory manual is a culmination of over 20 years of individual and group experiences. As with most general chemistry laboratory manuals, originality and fundamental concepts are blended to the point that most manuals are not so much unique as they bear the "flavor" of their institutions of origin. This being said, this manual finds its origins at North Carolina State University as "EXPERIMENTS WITH CHEMICAL REACTIONS," where it went through many revisions by Dr. Forrest Hentz and Dr. Gilbert Long. Dr. Andrew Bryan, a graduate of NCSU, performed many of the experiments as an undergraduate chemistry major and taught them as a graduate student. With permission, he began using them at Campbell University in 1990. From 1990 to 1998, they were modified by Dr. Bryan and Mrs. Ellie Luethy. Since 1998, Dr. Lin Coker has had the main responsibility of looking after these experiments and adding to them. Recently Mrs. Kimberly Elmore has joined in the process of enhancing this work. Dr. Coker and Mrs. Elmore would like to thank all of those who have helped in this work.

GENERAL CHEMISTRY LABORATORY

Safety Guidelines

1. Always wear splash-proof safety goggles in the chemistry laboratory. Know where the eyewash fountain is in your lab.

2. Always wear clothing consistent with safety; chemically resistant lab coats or aprons are recommended. Do not wear shorts, cutoffs, or miniskirts. Do not wear high-heeled shoes, open-toed shoes, sandals, or shoes made of woven material. Do not wear tank tops or midriff shirts. Confine long hair and loose clothing.

3. Never eat, drink, chew gum or tobacco, smoke, or apply cosmetics in areas where chemicals are used or stored.

4. Never perform any work when <u>alone</u> in the chemical laboratory. At least two people must be present. <u>Undergraduate students must be supervised by a lab instructor or trained lab assistant at all times.</u>

5. Never perform unauthorized work, preparations, or experiments.

6. Never engage in horseplay, pranks, or other acts of mischief in chemical work areas.

7. Never remove chemicals from the facility without proper authorization.

8. Consider all chemicals to be potentially hazardous.

9. Always dispose of chemicals, waste, and by-products in accordance with instructions from your laboratory instructor.

10. All injuries, accidents, and "near misses" must be reported to the lab instructor.

11. Visitors in a lab are required to abide by all safety guidelines and dress codes.

12. People with certain health conditions (including, but not limited to, pregnancy, breathing problems, or recent eye surgery) should consult their physician before taking a chemical laboratory course.

13. Always wash hands and arms with soap and water before leaving the work area. This applies even if one has been wearing gloves. (If wearing gloves, wash gloves before removing. Then, wash again.)

GENERAL CHEMISTRY DRAWER EQUIPMENT

Clay Triangle

Spatula

Scoopula

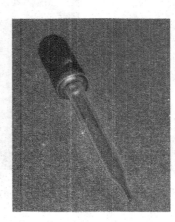

Dropper

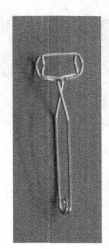

Test Tube Holder

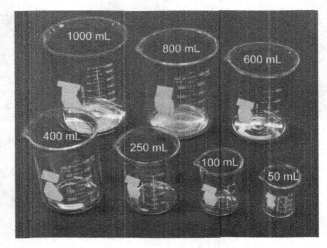

Beakers: 1000, 800, 600, 400,
250, 100, and 50 mL
Smaller beakers are nested inside larger
beakers for storage.

Glass Rods

Funnel

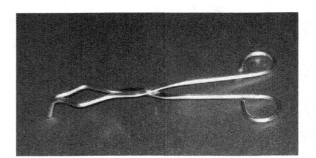

Tongs

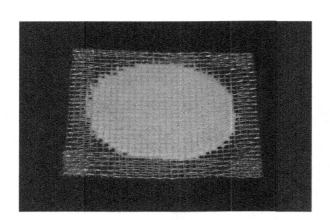

Wire Gauze

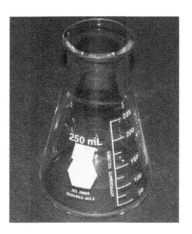

Three 250 mL Erlenmeyer Flasks

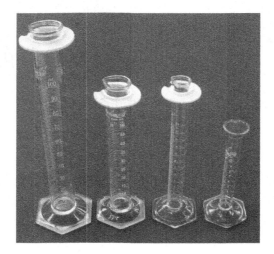

Graduated Cylinders: 10, 50, 25, and 100 mL
Graduated cylinders are placed flat for storage.

Cabinet Equipment For General Chemistry

Ring Stand

Bunsen Burner with Tubing

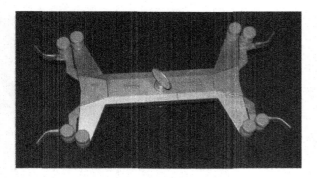

Burette Holder

Iron Ring

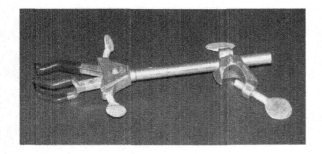

3-Prong Clamp with Holder

Safety Gas Lighter

EXPERIMENT 1: Introduction to the Chemistry Laboratory

PRELAB EXERCISE

Terms: *Define the terms below. You may have to refer to your text, a dictionary, or other reference.*

Accuracy –

Precision –

Meniscus –

Safety Warnings: *List any safety warnings described in the write-up or mentioned by your lab instructor.*

EXPERIMENT 1: Introduction to the Chemistry Laboratory

PRELAB EXERCISE

1. Which glassware can be used for accurate measurement?

2. Which glassware can be used for approximate measurement?

3. Which glassware can be used for estimated measurement, as a container, or for transfer?

4. Given a piece of glassware has markings to the tenths (X.X) place, to what place is a measurement recorded?

5. Describe the proper technique for reading a meniscus.

6. Describe the appearances of hot and cool flames.

INTRODUCTION TO THE CHEMISTRY LABORATORY

Objectives

1. To learn the proper technique of volume measurement, including a comparison of different types of glassware for volume measurement.

2. To learn the use of top-loading balances for mass measurements.

3. To learn proper technique for Bunsen burner use.

1.1 INTRODUCTION

Measurement or quantification is an important part of research. In this lab you will be learning to use various tools and techniques of the chemical laboratory. This lab is limited to volumetric measurements of liquids and the use of the electronic balance for measurements of mass. The additional technique of proper use of the Bunsen burner will be introduced.

Glassware

The volume measurement of liquids can be achieved with different types of glassware. Graduated cylinders are used to measure approximate variable volumes of liquids. Beakers are only used for estimated measurements, as a container, or for transfer. Volumetric containers (i.e., volumetric flasks and volumetric pipets) and burettes are used when accurate measurements of liquids are necessary. Volumetric containers are for exact volumes. Burettes are used for variable volumes.

Accuracy and Precision

There are differences in accuracy dependent upon the glassware used for measurement. The slightest contamination in glassware affects the accuracy of measurement. It is therefore very important that all glassware be clean when used. Volumetric glassware has been calibrated at a given temperature so differences in temperature can affect volume measurements. (Reminder: _Accuracy refers to how close the measurement comes to the true value. Precision refers to the degree of reproducibility of a measurement._ Therefore a set of values may be very precise but not accurate.)

Glassware Markings

There is one etched marking on volumetric glassware. When glassware has multiple markings, volume should be estimated to one place beyond the marked divisions of the glassware.

Reading a Meniscus

The liquid surface in a glass tube will be curved, either concave (downward) or convex (upward). This curved liquid is called a *meniscus*. Your eye <u>must</u> be level with the meniscus to obtain an accurate volume reading. You read from the bottom of the meniscus.

Electronic Balances

Electronic balances are used to determine mass. A typical electronic balance can be read to TWO digits beyond the decimal point—for example, a sample weighing a little more than 10 g might be measured to weigh 10.10 g. When multiple measurements are required, it is important to use the same balance to reduce error.

Pipets

There are two types of pipets, measuring and volumetric. A serological pipet is a specific type of measuring pipet. There are calibrated units etched on serological pipets. A volume may be estimated to one place beyond these units. Volumetric pipets are used to deliver a fixed volume of liquid.

Either type of pipet may be marked **TD** or **TC**. A pipet marked **TD**, short for "to deliver," means the pipet will deliver the specified volume if allowed to drain naturally. A pipet marked **TC**, short for "to contain," means that the pipet will contain the specified volume. Any amount of liquid left in the pipet should be gently blown out using the pipet bulb. (***However, at no time should your mouth EVER contact the pipet.***)

In this experiment you will use the volumetric pipet. A volumetric pipet is designed to accurately deliver a specified amount of liquid. A volumetric pipet is accurate to two decimal places. For example, a 15 mL volumetric pipet should be read as 15.00 mL.

Readings

Any readings from a scaled device (like a ruler) should be taken to ONE place beyond the markings on the device. All readings must include the units of the measurement.

Bunsen Burner Technique

A Bunsen burner is used to provide a quick, fairly hot, heat source. Air is mixed with fuel (methane or natural gas) and ignited. In the proper portions the reaction is as follows:

$$CH_4(g) + 2O_2(g) \rightarrow CO_2(g) + 2H_2O(l)$$

The air supplying oxygen for the reaction is brought into the burner through the base of the burner barrel. A correct mixture of fuel and air will produce a blue flame with a lighter blue cone in the center of the flame. The tip of the inner cone is the hottest point of the flame. In a properly adjusted flame you may hear a rustling sound.

If the air and fuel mixture has too little oxygen, the flame may appear yellow and produce soot and smoke due to the unburned carbon particles in the flame. The temperature of this type of flame is lower than in a blue flame.

If there is fuel coming into the burner faster than it can burn, you may see any of the following: burning above the barrel, failure to light, or a flame that continues to blow out.

The most useful flame burns relatively quietly and has two distinct zones; the inner one is cone-shaped and light blue. The height and nature of the flame should be adjusted to suit the nature of the job.

_**BUNSEN BURNER SAFETY WARNING: SHOULD THE FLAME NOT LIGHT
OR GO OUT, OR SOMETHING UNEXPECTED HAPPENS, TURN OFF THE GAS
AT THE GAS SUPPLY VALVE ON THE BENCHTOP! MAKE SURE THAT YOU
TURN OFF THE GAS AT THE BENCHTOP GAS SUPPLY VALVE WHEN YOU NO
LONGER NEED THE FLAME.**_

Burettes

Burettes are used to accurately deliver variable amounts of liquid. The markings on a burette are read in milliliters. A burette is marked to 0.X mL and should be read to 0.XX mL.

1.2 EXAMPLE CALCULATIONS

Note: In question 1 below, theoretical results, T, are the expected or reference results.
Experimental results, E, are your own determined results. In question 2, the accepted
value, A, refers to a value that is believed to be true and without error. The determined
value, D, is the value you find experimentally.

1. When a beaker is read to the 20 mL mark, it is found to deliver 21.99 mL. What is
    the percent difference?

$$\frac{|T - E|}{\left(\frac{T + E}{2}\right)} \times 100\% = \frac{|20 - 21.99|}{\left(\frac{20 + 21.99}{2}\right)} \times 100\% = 9.48\% = 9\%$$

2. A 50 mL burette is incorrectly read from the bottom instead of the top. The student
    wrote down 48.50 mL when the reading was actually 1.50 mL. What is the percent
    error caused by this mistake?

$$\frac{|A - D|}{A} \times 100\% = \frac{|1.50 - 48.50|}{1.50} \times 100\% = 3130\%$$

1.3 EXPERIMENTAL

Record all answers and observations on the appropriate data sheet. Remember that measurements must include units.

Part A: Precision Comparison of the Graduated Cylinder to the Beaker

Beaker   Graduated Cylinder

The precision of volumes measured by glassware in this lab can be markedly different. Use tap water and a 50 mL graduated cylinder for the following measurements. Remember to estimate ONE decimal place beyond the gradations on the graduated cylinder.

Fill a 50 mL beaker to just under the 10 mL mark; now, using a dropper, add water until the meniscus bottom is on the 10 mL mark. Remember, to read glassware as accurately as possible you must be eye level with the meniscus bottom when filling. Due to the gradations on the beaker, the reading would be to the nearest milliliter. Empty the beaker into your graduated cylinder and record the volume indicated on the graduated cylinder. Due to the gradations on the graduated cylinder, the reading should be to the nearest tenth of a milliliter. Shake the cylinder as dry as possible and repeat the measurement two additional times. Repeat the process (three times each) for 20 mL and 40 mL. Determine the average volumes for each of the three sets of measurements. Use the requested beaker volume as the reference to calculate the average percent difference in the volume delivered for each of the three different volumes.

Part B: Precision Comparison of the Graduated Cylinder to the Volumetric Pipet

Volumetric Pipet

Refer to "Appendix C: Volumetric Pipet" for directions on the use of a volumetric pipet.

In this step you will be using deionized water as the liquid that is being transferred. Use the volumetric pipet to transfer the specified volume (e.g., if you have a 10 mL pipet you are transferring 10.00 mL) of deionized water to a clean, dry 50 mL graduated cylinder. Record the pipet volume and the corresponding graduated cylinder volume reading to the correct number of significant figures. Empty the graduated cylinder and gently shake dry. Repeat this process for a total of three trials.

Determine the average values for the pipet and graduated cylinder volumes. Using the average values, determine the percent difference between the pipet and graduated cylinder volumes. You will use the pipet volume as the reference value.

Note: You earn five points for successfully demonstrating this technique to your instructor or the assistant and having one of them sign off on your proficiency.

Part C: Precision Comparison of the Graduated Cylinder to the Burette

Refer to "Appendix D: Burette" for instruction on how to set up a burette.

Set up the burette as described and fill it to between the 0.00 and the 1.00 mL mark with tap water. (Remember to check around the bottom

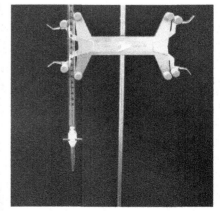

Burette

of the stopcock and in the burette tip for air pockets. If there is an air pocket, drain the burette just until the air pocket is flushed from the burette.) Due to the graduations on the burette, the reading would be to the nearest hundredths place. Record the initial reading.

Drain 10–11 mL of liquid from the burette into a clean 50 mL graduated cylinder. Record the final reading for the burette and the corresponding graduated cylinder volume to the correct number of significant figures. Empty the graduated cylinder and gently shake dry. Repeat this process for a total of three trials.

Use the requested burette volume as the reference to calculate the percent difference for the averages in the volume delivered. Demonstrate the proper use of the burette to your instructor or assistant.

Note: You earn five points for successfully demonstrating this technique to your instructor or the assistant and having one of them sign off on your proficiency.

Part D: Mass Measurements

Electronic Balance

Refer to "Appendix A: Electronic Balance" for directions on the use of the electronic balance.

Like many objects, each tablet has a slightly different mass. To determine the mass of a typical tablet it is easier to determine the mass of a known number of tablets and determine the average. For this section obtain 3 tablets and properly measure the mass of the group of 3 tablets. Using this information calculate the mass of a typical tablet. Demonstrate proper balance technique to your instructor or assistant.

Note: **Never return unused portions of a chemical to its storage container**. By doing so you could contaminate the container. A waste container for the tablets will be provided.

Note: You earn five points for successfully demonstrating this technique to your instructor or the assistant and having one of them sign off on your proficiency.

Part E: Bunsen Burner Technique

Refer to "Appendix B: Bunsen Burner" for directions on the use of the Bunsen burner.

Practice lighting a flame in the Bunsen burner until you are able to properly light the flame, adjust for a hot flame and cool flame, adjust the flame height, and properly turn off the flame. Demonstrate for your instructor.

Note: You earn five points for successfully demonstrating this technique to your instructor or the assistant and having one of them sign off on your proficiency.

Bunsen Burner

Safety Notes

Be cautious using the Bunsen burner.

1.4 WASTE DISPOSAL

Waste for this lab may be flushed down the sink. The tablets go in the labeled waste container.

EXPERIMENT 1: Introduction to the Chemistry Laboratory

DATA SHEET 1

A. Precision Comparison of the Graduated Cylinder to the Beaker

Beaker Volumes   Graduated Cylinder Volumes

    10 mL   _____

    10 mL   _____

    10 mL   _____

Avg. Volume for 10 mL Graduated Cylinder Measurements: _____

% Difference for 10 mL Avg. Volume: _____

Beaker Volumes   Graduated Cylinder Volumes

    20 mL   _____

    20 mL   _____

    20 mL   _____

Avg. Volume for 20 mL Graduated Cylinder Measurements: _____

% Difference for 20 mL Avg. Volume: _____

Beaker Volumes   Graduated Cylinder Volumes

    40 mL   _____

    40 mL   _____

    40 mL   _____

Avg. Volume for 40 mL Graduated Cylinder Measurements: _____

% Difference for 40 mL Avg. Volume: _____

EXPERIMENT 1: Introduction to the Chemistry Laboratory

DATA SHEET 2

B. Precision Comparison of the Graduated Cylinder to the Volumetric Pipet

Pipet Volume _____   Graduated Cylinder Volume _____

Pipet Volume _____   Graduated Cylinder Volume _____

Pipet Volume _____   Graduated Cylinder Volume _____

Avg. Volume _____   **Avg. Volume** _____

% Difference   _____

Signature of lab instructor or lab assistant for correct pipet technique:

   Notes (if any):

C. Precision Comparison of the Graduated Cylinder to the Burette

   Burette Readings   Graduated Cylinder Volumes

   Initial   _____

   Final   _____

   Volume Delivered   _____   _____

   Initial   _____

   Final   _____

   Volume Delivered   _____   _____

   Initial   _____

   Final   _____

   Volume Delivered   _____   _____

   Avg. Volume   _____   _____

% Difference: _____

Signature of lab instructor or lab assistant for correct burette use:

   Notes (if any):

EXPERIMENT 1: Introduction to the Chemistry Laboratory

DATA SHEET 3

D. Mass Measurements

Name Brand and Type of Tablet   _____

Mass of 3 Tablets   _____

Calculated Avg. Mass of a Tablet   _____

Signature of lab instructor or lab assistant for correct balance use:

      Notes (if any):

E. Bunsen Burner Technique

Signature of lab instructor or lab assistant for safe Bunsen burner technique:

      Notes (if any):

EXPERIMENT 1: Introduction to the Chemistry Laboratory

POSTLAB EXERCISE

1. How would you correct a Bunsen burner flame when:

    a. the flame is burning above the barrel?

    b. the flame appears yellow and produces soot and smoke?

2. Briefly describe why the accuracy of the values read from a beaker is less than that from a graduated cylinder.

3. A _____ mL volumetric pipet is determined to actually deliver _____ mL. What is the percent difference?

4. A burette was improperly read to one decimal place giving a reading of _____ mL. If the actual volume is _____ mL, what is the percent error caused by this mistake?

EXPERIMENT 2: Density of Solids and Liquids

PRELAB EXERCISE

Terms:

    Volume by displacement –

    "Best straight line" –

    Density –

Safety Warnings:

EXPERIMENT 2: Density of Solids and Liquids

PRELAB EXERCISE

1. What is the density of an object with a volume of 8.29 mL and a mass of 16.31 g?

2. What are the three steps to determine the density of an irregularly shaped object?

3. A cube has a mass of 501.00 g and a density of 5.739 g/cm^3. What is the length of a side of this cube?

4. A cylinder has a density of 3.75 g/cm^3. If it has a mass of 12.1 g and a diameter of 8.22 cm, what is its height?

DENSITY OF SOLIDS AND LIQUIDS

Objectives

1. To determine volume by direct measurement.

2. To determine volume by indirect measurement (i.e., "volume by displacement").

3. To gain experience using top-loading balances for mass measurements.

4. To determine density by calculation.

5. To determine density through slope of a line.

6. To generate a graph correctly.

2.1 INTRODUCTION

Densities of liquids can be determined simply by measuring volume with an appropriate piece of glassware and determining the mass using a balance.

Determination of the volume of an object can be made two ways. First, the volume of an object can be determined by direct measurement if the object has a regular shape. Using a centimeter ruler, for example, lengths, widths, heights, or diameters can be estimated. Volume is then determined by knowing the correct formula for volume. The following are volume formulas to use when direct measurements are possible.

Volume of a Cube   Length^3

Volume of a Box   Length $\times$ Width $\times$ Height

Volume of a Cylinder $\pi r^2 \times$ Height

Volume of a Sphere   4/3 πr^3

(Note: r = Radius = 1/2 $\times$ Diameter)

Second, if an object is irregularly shaped, volume can be determined by *displacement* using a graduated cylinder. Fill a graduated cylinder to a volume that will at least cover the object(s) that will be added to the cylinder. Take an initial reading, add the object(s), and take a final volume reading. The difference between the final and initial volumes will be the displaced volume. (Note: It is desirable to get a displacement of at least 5 mL in a 25 mL graduated cylinder or a 10 mL displacement in a 50 mL graduated cylinder to achieve a fairly accurate density. Use a cylinder that offers the most divisions for the volume you are trying to measure.)

Once the mass and volume of your object(s) or liquid are determined, density can be determined by the following:

$$\text{Density of Object} = \frac{\text{Mass of Object}}{\text{Volume of Object}}$$

When several different corresponding mass and volume measurements are taken, determining the slope of a graph of mass versus volume will give a better estimate of the density (if mass is plotted versus volume, the slope has units of g/mL, which is density). This will take all your measurements into account and reduce the impact of incorrect measurements. You will use a computer spreadsheet to assist you in graphing your data.

2.2 Example Calculations

1. An object with a mass of 5.43 g has a volume of 3.37 cm^3. What is the density of the object in g/cm^3?

$$\frac{5.43 \text{ g}}{3.37 \text{ cm}^3} = 1.61 \text{ g/cm}^3$$

2. A cylinder has a radius of 3.01 cm and a height of 7.74 cm. It has a mass of 193.89 g. What is the density of the object in g/cm^3?

$$V_{cylinder} = \pi r^2 h = (3.1416)(3.01 \text{ cm})^2 (7.74 \text{ cm}) = 220.3 \text{ cm}^3$$

$$d = m/v = 193.89 \text{ g}/220.3 \text{ cm}^3 = 0.880 \text{ g/cm}^3$$

Remember to show your work for calculations.

2.3 Experimental

Part A: Determining the Density of a Regularly Shaped Object

Your lab instructor will assign you a wooden block. Record the identity of the block. Determine the dimensions of the block using a ruler and record your results. Remember to *estimate one place past the markings on your ruler*. Now measure and record the mass of the block. Using the digital balance, first place a piece of weighing paper on the balance and press the tare or "re-zero" bar; this should cause the balance to read "0.00 g." If not, press it again! (Note: Remember to use weighing paper.) Return the wooden block to your instructor.

Obtain and record the information of three additional blocks from your classmates. You will also need the names of the classmates you are sharing this data with.

Calculate and record the volume of the blocks in cubic centimeters. Calculate and record the density of your block using an appropriate formula. Remember to use an appropriate number of significant figures.

Part B: Determining the Density of a Small or Irregularly Shaped Object

Before beginning, obtain a sample of a particular metal with a volume of at least 10 mL but as close to 10 mL as possible. Each sample will be labeled (e.g., L or Z); record the label.

For your sample, place enough water in a 50 mL graduated cylinder so that the water will completely cover your metal sample when it is added to the cylinder. A larger graduated cylinder may be needed for longer metal cylinders. Record the volume as Volume$_i$. Tilt the cylinder and allow your metal sample to *gently slide* to the bottom of the cylinder. Record the volume as Volume$_f$. Dry your sample then determine the mass of your sample and record. If your sample is metal spheres, measure the mass in a tared beaker. Remember to use weighing paper. Metal cylinders can be weighed directly on the weigh paper. Repeat this process to get three estimates of your sample's volume and mass.

Return the metal pieces to the side-shelf after drying them with a paper towel.

Your lab instructor will provide you with the accepted value for the density of the metal you used in the experiment. Use this value to calculate the percent error for the density of your sample.

Part C: Density of a Liquid

The following, while allowing the determination of the density of a liquid, shows that a *linear relationship exists between the mass and volume of a substance.*

Carefully clean and dry a 100 mL graduated cylinder and a funnel. Place some weigh paper on the balance and auto-zero the readout. Weigh the graduated cylinder with the funnel on top of the graduated cylinder to the nearest 0.01 gram. Record this mass. Leave the graduated cylinder and funnel on the balance. Make sure the funnel is set so that any liquid going through it will go to the bottom of the graduated cylinder. Pour some of the unknown liquid into a beaker. Add 8–9 mL from the beaker into the graduated cylinder. Read the volume to the correct number of significant figures. Determine the mass of the graduated cylinder to the nearest 0.01 gram. Record these data on your data sheet.

Without emptying the graduated cylinder, add an additional 8–9 mL of your liquid, re-weigh, and record your data. Repeat this procedure until you have made a total of five different liquid additions to the graduated cylinder.

Place the liquid in the provided waste container.

Safety Notes

There are no unusual hazards in this experiment.

2.4 WASTE DISPOSAL

Return the wooden blocks to your lab instructor. Return the metal pieces to the side-shelf. Place the liquid in the provided waste container.

2.5 GRAPHING

Refer to "Appendix E: Generating a Straight Line Graph in Microsoft Excel" for directions on graphing.

Construct a graph of mass of the graduated cylinder, funnel and liquid added versus the volume of the liquid. You are required to use the computer to generate this graph. Your lab instructor will help you create the graph during lab. You can also get help during your lab instructor's office hours.

EXPERIMENT 2: Density of Solids and Liquids

DATA SHEET 1

A. Density of a Wooden Block

Identity of Block				
Mass of Block				
Height				
Width				
Length				
Volume (H x W x L)				
Density				
Name of Student Performing Measurement				

B. Density of Small or Irregularly Shaped Objects

Trial	Mass of Metal	Volume$_i$	Volume$_f$	Volume$_{calc}$ (at least 10 mL)	Density
1					
2					
3					

Code of Metal _____

Actual Density of Metal (from lab instructor) _____

Avg. Determined Density of Metal _____

% Error _____

EXPERIMENT 2: Density of Solids and Liquids

DATA SHEET 2

C. Density of Liquid

	Initial	1	2	3	4	5
Mass: Graduated Cylinder Funnel & Liquid						
Total Volume of Liquid	0					

Identity of Liquid _____

Density of Liquid _____

Y-**Intercept** _____

EXPERIMENT 2: Density of Solids and Liquids

POSTLAB EXERCISE

1. Based upon your results, if the densities in Part A have a 1% uncertainty associated with them, what would be the density range for each of the four wood samples?

2. If the densities of the four wood samples in Part A have a 1% uncertainty associated with them, can you distinguish between your block of wood and the other three samples using only the density ranges? Explain your reasoning.

3. Object _____ is less dense than Object _____. If both objects are the same mass, what can be said about the *volume* of A as compared to the *volume* of B? Explain your reasoning.

4. A spherical-shaped object has a diameter of _____ cm and a mass of _____ g. What is its density in g/cm^3?

EXPERIMENT 3: Properties of Household Chemicals

PRELAB EXERCISE

Terms:

    Chemical property –

    Physical property –

    Solubility –

    Malleability –

    Conductivity –

    Electrolyte –

    pH –

    Starch test –

Safety Warnings:

EXPERIMENT 3: Properties of Household Chemicals

PRELAB EXERCISE

1. Why is it important to clean your spatula each time you use it to obtain a different household solid?

2. What has happened if a chemical reaction has occurred?

3. What two things are unaffected in a physical change?

4. List five indications of a chemical change.

5. What do pH and conductivity have in common?

PROPERTIES OF HOUSEHOLD CHEMICALS

Objectives

1. To experimentally observe differences in chemical properties by mixing household solids with household liquids.

2. To experimentally observe differences in physical properties through observation of such things as color, texture, malleability, and solubility.

3. To identify an unknown solid through similarities of chemical and physical properties with the other solids tested.

3.1 INTRODUCTION

In this experiment, you will be called on to make observations. These observations will either be of an existing property of a substance or the change a substance undergoes when exposed to another substance or condition (heat, light, etc.). Your observations of several white substances will be used to determine the composition of an "unknown" mixture of the white substances. Your observations will fall into two categories: (1) physical changes/properties and (2) chemical changes/properties.

In a **physical change** the appearance of a substance changes but its composition and identity are unaffected. Examples of physical changes include the boiling of water to produce steam, the filing of metal to produce dust or filings, and the dissolving of sugar in water to form sugar syrup. In most cases, a matter of one or two simple processes will reverse the physical change. For example, sugar may be isolated from sugar syrup merely by evaporating the water. **Physical properties** are such things as color, state (solid, liquid, or gas), density, and melting and boiling points. *They do not involve a chemical reaction.*

In a **chemical** reaction a **change** in the composition and identity of a substance occurs. Some chemical reactions are the burning of wood to form carbon dioxide and water, the rusting of iron to form iron oxide, and the heating of limestone to form lime and carbon dioxide. The substances present before a chemical reaction occurs are reactants. The substances formed are called the products and these *products have physical and chemical properties that are different from those of the reactants.* **Chemical properties** are displayed when a substance undergoes a chemical reaction to produce products. Reversing a chemical reaction usually requires an involved process of several steps. For example, green plants use photosynthesis, a complex series of reactions that science has not yet duplicated, to make wood from carbon dioxide and water. A chemical reaction is indicated by any of the following observations:

1. Change of color.

2. Production of heat, light, or sound.

3. Evolution of a gas.

4. Formation of a solid where none was present before.

5. Formation of ions (observable with meters, indicators, or test strips).

These observations are not infallible indications that a chemical reaction has occurred. For example, when ice forms from liquid water, a solid is present where none was before, but a physical change rather than a chemical reaction has occurred. In this case, one would classify the change as physical based on the ease of reversing it.

3.2 EXPERIMENTAL

For this experiment you need the following items at your lab station:

7 white solids (table salt, baking soda, sugar, citric acid, vitamin C, starch, baking powder) and an unknown

2 well plates	small spatula
small stirring rod	magnifying lens
8 small test tubes	test tube clamp or tongs
test tube holder	Bunsen burner
pHydrion paper	conductivity probe

Observation of Unknowns

Your lab instructor will assign you an "unknown" which contains one of the solids from above. They may also contain an unknown solid which may interfere with some of your tests. Perform the same tests (Parts A, B, C, D, and E) on the unknown at the same time as your knowns. This will make it easier to compare the results to determine which solid is in your unknown. Be sure to include the number of your "unknown" powder on your data sheets where requested along with your observations.

Note: Parts A, B, and C should be performed at the same time. Place a spatula tip of each powder into the two neighboring wells of the well plate. Do this for each known and your unknown. You may need to use two well plates for this.

Part A: Observations on Visible Characteristics of Solids

Carefully observe the physical properties of the powders. Use the magnifying glass and stirring rod to determine color, texture, and **malleability** (degree to which shaping is possible). When using the magnifying glass note whether or not the unknown has a crystalline structure, and if so, are the crystals irregularly or regularly shaped? If the crystals are regularly shaped, what shape are they? *Record your observations.*

Part B: Observations on Mixing Solids with a Household Acid

Test the top row of samples as to their reactivity with common household acid, vinegar (a.k.a. acetic acid, $HC_2H_3O_2$). Place 5 drops of vinegar on each solid. Note any chemical changes that may take place such as bubbling (formation of a gas; "G"), color changes ("C"), or generation of heat ("H"). (Determine heat generation by touching the bottom of the well.) Stir the mixture of solution and powder with a *clean* stirring rod or gently swirl the well plate to determine if the powder dissolves in the liquid. A sample dissolving in a liquid is an example of solubility ("S"). *Record all of your observations on the sheet provided.*

Part C: Specific Test for the Presence of Starch

Test the bottom row of samples for the presence of starch. Do this by adding 2 drops of the solution labeled "Iodine Solution." If starch is present, a blue-black color will develop. *Record all of your observations on the sheet provided.*

Part D: Observations on Dissolving the Solids in Water

Measuring Solubility

Place 0.5 g of each solid into a separate 50 mL beaker; add 40 mL of deionized water to each beaker and stir. Also, in a clean beaker, set aside about 20 mL of the water used to make the solutions. Make observations on solubility for the mixtures. Is the solution cloudy? Is the solid completely dissolved? Was heat liberated/absorbed?

Note: Use the solutions to perform the following tests on conductivity and pH.

Measuring Conductivity

Conductivity is the ability of a solution to conduct electricity; this property indicates the presence of ions. Generally speaking, conductivity increases as the concentration of a particular ion increases and as the magnitude of the charge on the ion increases. For example, two solutions with equal concentrations of Na^+ would have similar conductivities, but a similar concentration of Fe^{3+} would have a greater conductivity. Solids that dissolve without producing ions are called **nonelectrolytes** and are considered poor conductors. Those solids that dissolve and completely break into ions are called **strong electrolytes**. Other solids that have intermediate conductivity are called **weak electrolytes**.

To measure conductivity you will use a conductivity meter. Directions for calibration are provided in the box below: Specific instructions will be provided by your lab instructor.

Using the Vernier Conductivity Probe

1. Rinse electrode with deionized water and blot dry with a lint-free tissue.
2. Swirl or stir solution and place the electrode in the solution.
3. Allow time for reading to stabilize then record it with the unit.
4. Rinse electrode with deionized water and blot dry between readings.

Carefully record the conductivity and the displayed unit of each solution prepared in Part D. In the data sheet, try to categorize substances as nonelectrolytes, weak electrolytes, and strong electrolytes. Also measure the conductivity of the deionized water that was set aside for comparison. Deionized water is considered a nonelectrolyte.

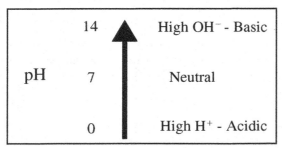

Measuring pH (acidic, basic, neutral)

Changes in the hydrogen ion (H^+) and hydroxide ion (OH^-) concentrations can occur when solids are dissolved in water. According to the Arrhenius definition of acids and bases, **acids increase the H^+ ions in aqueous solutions**, whereas **bases increase the OH^- ions in aqueous solutions**. Because of these characteristics, acids have low pH (0 to 7) and bases have high pH (7 to 14). This is summarized in the figure shown here.

The pH of a solution can be measured with meters, indicators, or test paper. In this experiment pH will be measured with pHydrion paper which changes color with pH.

Test each solution prepared for Part D. Tear off a small strip (approximately 1 cm) for each solution and place it on a paper towel. Take a clean stirring rod and dip it in your beaker and then touch it to a strip of the pHydrion paper. You should observe a color change. Record the pH corresponding to the color. Also test the deionized water that was set aside. Why is the pH of the "pure" water not observed to be neutral?

Part E: Observations on Heating the Solids

Refer to "Appendix B: Bunsen Burner" for directions on the use of the Bunsen burner. Place a small amount of each powder into separate test tubes. Using a test tube clamp and Bunsen burner, strongly heat the test tube with the sample using a blue flame with a lighter blue cone in the center of the flame. Place the test tube at the tip of the inner cone which is the hottest point of the flame. Note any changes that take place when the powders are heated. Does the appearance of the powder change? If so, how would you describe the texture? Is there a color change? Is a gas given off? Is there a characteristic odor? These are examples of changes that *may* occur. You may note other changes or no changes. *Record your observations on the sheet provided.*

Note: *Place hot test tubes into a small <u>beaker</u> to allow them to cool.* If the powder undergoes a combustion reaction, the test tube will have to be placed in the glass discard container *after being allowed to cool.* The other test tubes can be rinsed out and returned to their original location for reuse.

Safety Notes

Use the Bunsen burner with caution. The chemicals used in this experiment are commonly found in homes throughout the country. To gain experience, treat these chemicals with care.

3.3 Waste Disposal

All test tubes should be rinsed with tap water. Test tubes that appear clean after this may be returned to their original location for reuse. Test tubes that appear dirty after this should be allowed to cool and then placed into a glass discard container. Remaining waste for this lab may be dissolved in tap water and flushed down the sink.

NAME: _____ LAB SECTION: _____

EXPERIMENT 3: Properties of Household Chemicals

DATA SHEET

Record your unknown number _____

Sample	Salt	Sugar	Baking Soda	Citric Acid	Starch	Vitamin C	Baking Powder	Unknown
Color								
Texture								
Malleability								
Other Visual Notes								
Acid Reaction								
Starch Test								
Solubility								
Conductivity								
pH								
Heating								

Conductivity of deionized water _____

pH of deionized water _____

EXPERIMENT 3: Properties of Household Chemicals

POSTLAB EXERCISE

1. Why is it important to check the conductivity and pH of "pure water" for comparison?

2. Based on the results from the experiment, what is the solid in your unknown (5 points)?

3. Summarize your reasoning for your answer to question 2 (5 points).

   Color/Texture/Malleability/Visual Notes

   Acid Reaction

   Starch Test

   Solubility

   Conductivity

   pH

   Heating

EXPERIMENT 4: Formation and Naming of Ionic Compounds

PRELAB EXERCISE

Terms:

Cation –

Anion –

Precipitate (S) –

NR –

G –

Safety Warnings:

EXPERIMENT 4: Formation and Naming of Ionic Compounds

PRELAB EXERCISE

1. List two visible changes that may occur in this experiment when a cation and anion are mixed together.

2. Which abbreviation will you use to indicate there was no visible change when a cation and anion were mixed?

3. How many drops of anion are placed into a well in the well plate?

4. List the cations used in this experiment.

FORMATION AND NAMING OF IONIC COMPOUNDS

4

Objectives

1. To experimentally cause the formation of an ionic compound by mixing a cation with an anion.

2. To properly write the formula for an ionic compound.

3. To properly name an ionic compound.

4. To correctly identify an "unknown" anion by comparing the properties of the unknown to the properties of the "known" anions.

4.1 INTRODUCTION

When solutions of cations and anions are mixed, a chemical reaction can occur. Some reactions result in a visible change such as the evolution of a gas or the formation of a precipitate (solid). Although not visible, there may be an acid–base change if a reaction occurs between an anion and cation. Sometimes no visible changes occur because no reactions are occurring.

In this lab you will be making observations upon mixing a series of cation/anion pairs. You will identify the reactions as the following:

1. Cation + Anion → Precipitate (S)

2. Cation + Anion → Gas (G)

3. Cation + Anion → No Reaction (NR)

Note: In this experiment you will only check for gas or precipitate formation as evidence of a reaction between cations and anions. You are required to check the pH of the anion only as a help to identify your unknown anion.

For the postlab, you will be asked to write a tentative formula for several cation/anion pairs based on the ion charges. (Note: This may not be the true identity of the compound, which is formed, but it is a good first guess based on your current knowledge of solution chemistry.)

All the data you collect will be useful to you, since your observations will be used to determine the identity of the unknown anion assigned to you.

4.2 EXPERIMENTAL

The table in the data sheet shows the ion solutions you will be using in this experiment. Be careful in all observations. If you are **unsure** of a result, **repeat it** but strive not to waste reagents.

For this experiment you will be using two well plates, each containing 24 small reaction wells. Thoroughly clean the plates and rinse with *deionized* water. Remove excess deionized water by vigorously shaking the well plates a few times. (Drying with a paper towel is not necessary and will leave lint in the wells.) Place the plates on your bench in horizontal orientation, one above the other. Together the plates will have six columns and eight rows. To ease identification, the columns are labeled with a numeral from 1 to 6, and the rows are labeled with a letter from A to D.

Starting in well 1A and going down the first column, place 5 drops of Cl^- solution into each of the first five wells. Continue in this manner with the remaining known anion solutions and your unknown, starting at the top of a different column for each anion and working in the order given in the table in your data sheet.

Starting in well 1A and going across the first row, add 1 drop of H^+ solution to each of the six wells and *gently* swirl to mix, being careful not to cross-contaminate solutions. Inspect each well for a reaction. If no reaction is evident in a well, add a second drop of the H^+ solution, swirl, and re-inspect. Record your observations in the data sheet table. For precipitates, record **S** and the color. (Haziness may constitute a precipitate if it settles to the bottom.) For gases that evolve (bubbling), record a **G**. When no reaction is observed (no solid or no gas), record **NR**. Continue in this manner with the remaining cation solutions, starting with a different row for each cation and working in the order given in the table.

Obtain two strips of pH test paper and tear each strip into three pieces; place the pieces on a paper towel. Place a drop of the Cl^- solution on the first piece of pH paper then use the chart on the test paper container to determine the pH that corresponds to the color of the wetted paper. Immediately record your results in the table in your data sheet. Repeat the test with the remaining anions and your unknown, using fresh pH test paper for each anion and in the order given in the table. (The cations are not being tested for pH.)

Be sure to record the ID number for your unknown and what you have determined it to be. If you are not sure of the identity of your unknown, repeat the tests on your unknown.

Cross-contamination of reagents could negatively affect your results. If you feel cross-contamination may have occurred during the procedure, or if you are unsure of a result, you should repeat the procedure for any mixtures in question.

Safety Notes

The source of the H^+ ions is 3 M HCl, which can cause mild burns, and would be especially harmful to the eyes. The other reagents are approximately 0.2 M concentrations and are fairly mild irritants.

4.3 WASTE DISPOSAL

Waste for this lab may be flushed down the sink with copious amounts of tap water. The pH paper can go in the regular trash.

EXPERIMENT 4: Formation and Naming of Ionic Compounds

DATA SHEET

Data and Observations

Abbreviations: NR (no reaction), G (gas evolved), and S (precipitate forms). Where precipitates are formed, indicate the color observed.

Record your unknown number _____

	Cl^-	OH^-	CO_3^{2-}	SO_4^{2-}	PO_4^{3-}	Unknown
H^+						
Al^{3+}						
Ca^{2+}						
Cu^{2+}						
Fe^{3+}						

pH Paper Test	Cl^-	OH^-	CO_3^{2-}	SO_4^{2-}	PO_4^{3-}	Unknown
pH						

Determined identity of your "unknown" anion (5 points) _____

Explain your reasoning below: (5 points)

EXPERIMENT 4: Formation and Naming of Ionic Compounds

POSTLAB EXERCISE

For each of your cation/anion pairs record an appropriate name and chemical formula below: (1/2 point per blank)

	Name	**Formula**

Br^-

    Na^+ _____ _____

    Mg^{2+} _____ _____

    Ca^{2+} _____ _____

    Al^{3+} _____ _____

    Fe^{3+} _____ _____

NO_3^-

    Na^+ _____ _____

    Mg^{2+} _____ _____

    Ca^{2+} _____ _____

    Al^{3+} _____ _____

    Fe^{3+} _____ _____

Instructor provided anion:

    Na^+ _____ _____

    Mg^{2+} _____ _____

    Ca^{2+} _____ _____

    Al^{3+} _____ _____

    Fe^{3+} _____ _____

Instructor provided anion:

    Na^+ _____ _____

    Mg^{2+} _____ _____

    Ca^{2+} _____ _____

    Al^{3+} _____ _____

    Fe^{3+} _____ _____

Instructor provided anion:

    Na^+ _____ _____

    Mg^{2+} _____ _____

    Ca^{2+} _____ _____

    Al^{3+} _____ _____

    Fe^{3+} _____ _____

EXPERIMENT 5: Water Loss by Dehydration

PRELAB EXERCISE

Terms:

     Hydrate –

     Dehydration –

     Anhydrate –

     Crucible –

Safety Warnings:

EXPERIMENT 5: Water Loss by Dehydration

PRELAB EXERCISE

1. What is the difference in appearance between a hot crucible and a cold crucible?

Use the following information to answer the questions below:

A sample of $Mg(ClO_4)_2 \cdot XH_2O$ is analyzed using a procedure similar to that described in this experiment to determine its formula.

Mass of empty crucible and lid	93.70 g
Mass of crucible and lid and sample	99.91 g
Mass of crucible and lid and heated sample	97.79 g

2. What mass of hydrate is used in this experiment?

3. What is the mass of water of hydration in this experiment?

4. What is the mass of the anhydrous salt?

5. How many moles of water of hydration are there?

6. How many moles of anhydrous salt are there in this experiment?

WATER LOSS BY DEHYDRATION

Objectives

1. To use a Bunsen burner and crucible to dry a sample.

2. To experimentally determine the amount of water lost through a process known as "dehydration."

3. To calculate a mole ratio from grams remaining after dehydration.

5.1 INTRODUCTION

Certain salts called "hydrates" contain water molecules as part of their structure. (Hydrates are especially common among the salts of transition metals.) Water molecules are bound to the salts in definite stoichiometric proportions. The number of waters bound to a particular metallic salt is a characteristic of that compound. Three common hydrates are listed below:

Name	Common Name	Formula
Copper (II) sulfate pentahydrate	Blue vitriol	$CuSO_4 \cdot 5H_2O$
Calcium sulfate dihydrate	Gypsum	$CaSO_4 \cdot 2H_2O$
Magnesium sulfate ??? hydrate	Epsom salt	$MgSO_4 \cdot XH_2O$

(you will determine)

The name and formula of a hydrate salt is based on the salt and the number of water molecules attached. (For copper (II) sulfate pentahydrate, copper (II) sulfate is the salt's name; the prefix "penta" stands for 5; and hydrate represents water. In the formula, $CuSO_4$ is the formula for the salt, and the $\cdot 5H_2O$ represents the five attached water molecules.)

In a process called dehydration, many hydrates will give up their water molecules during heating, leaving what is called an anhydrous (without water) salt. For the case of copper (II) sulfate pentahydrate, one can say that five molecules of water could be driven off per one formula unit of salt, or that five moles of water can be driven off from one mole of salt.

Therefore, the following relationship exists:

$$\frac{5 \text{ moles } H_2O}{1 \text{ mole } CuSO_4 \cdot 5H_2O} \quad \text{or} \quad \frac{90.0 \text{ grams of } H_2O}{249.68 \text{ grams of } CuSO_4 \cdot 5H_2O}$$

In this experiment, you will use heating and weight loss techniques to determine the number of water molecules in a formula unit of Epsom salt.

5.2 EXAMPLE CALCULATIONS

A sample of $Mg(ClO_4)_2 \cdot XH_2O$ is analyzed using a procedure similar to that described in this experiment to determine its formula.

Mass of empty crucible and lid	85.34 g
Mass of crucible and lid and sample	92.68 g
Mass of crucible and lid and heated sample	90.29 g

1. What mass of hydrate is used in this experiment?

Mass of crucible and lid and hydrate	92.68 g
– Mass of empty crucible and lid	85.34 g
Mass of hydrate	7.34 g

2. What is the mass of water of hydration in this experiment?

Mass of crucible and lid and hydrate	92.68 g
– Mass of crucible and lid and anhydrous salt	90.29 g
Mass of water of hydration	2.39 g

3. What is the mass of the anhydrous salt?

Mass of crucible and lid and anhydrous salt	90.29 g
– Mass of empty crucible and lid	85.34 g
Mass of anhydrous salt	4.95 g

4. How many moles of water of hydration are there?

$$2.39 \text{ g } H_2O \times \frac{\text{mol } H_2O}{18.02 \text{ g } H_2O} = 0.132_6 \text{ mol } H_2O$$

5. How many moles of anhydrous salt are there in this experiment?

$$4.95 \text{ g Mg(ClO}_4)_2 \times \frac{\text{mol Mg(ClO}_4)_2}{223.21 \text{ g Mg(ClO}_4)_2} = 0.0221_8 \text{ mol Mg(ClO}_4)_2$$

6. What is the ratio of moles of water to moles of anhydrous salt?

$$\frac{0.132_6 \text{ mol H}_2\text{O}}{0.0221_8 \text{ mol Mg(ClO}_4)_2} = \frac{6.00 \text{ mol H}_2\text{O}}{\text{mol Mg(ClO}_4)_2}$$

7. What is the formula of this hydrate?

$$\text{Mg(ClO}_4)_2 \cdot 6\text{H}_2\text{O}$$

5.3 EXPERIMENTAL

Note: Careful research involves repeating an experiment. This is the reason for performing three trials.

Preparation of Crucible

Prepare a clean, dry crucible. Place the crucible and lid (have lid slightly opened on the crucible) on a clay triangle which has been placed on an iron ring attached to a stand. This is shown in Figure 1.

Figure 1 Crucible Setup

Heat the bottom of the crucible with a Bunsen burner until the bottom of the crucible is maintained red-hot for at least 5 minutes. Using your tongs, carefully transfer the lid and crucible to wire gauze and let cool. *Note: Placing the hot crucible on the cold counter could cause the crucible to crack. Do not use a piece of wood for the crucible to cool on. The wood will scorch. Remember that a hot crucible looks like a cold crucible.* When the crucible reaches room temperature, weigh the crucible and its lid and record the mass.

Removing the Water of Hydration of Epsom Salt

Prepare a clean, dry crucible as directed above. Weigh into the crucible between 2.00 and 2.50 g of Epsom salt and record this mass. *(When given a range of measurements, aim for the minimum.)* Heat by placing the crucible and its lid onto the clay triangle, arranging the lid so that it is slightly ajar, and begin heating the crucible with a very small flame. Continue heating with a small flame for about 5 minutes. Beware of spattering during this time. If spattering occurs, quickly push the lid over the crucible and remove the flame. Spattering is undesirable because it will increase the chances of sample loss.

After you have heated the crucible gently, increase the size of the flame somewhat, and heat the crucible and contents for 5 minutes after the bottom begins to glow. The lid is slightly ajar during this step.

Turn off the burner and remove your crucible from the tripod, placing it on your wire gauze. Allow the crucible to cool to room temperature, with the cover in place. After your crucible has cooled to the point that the crucible can be held in your hand, weigh it (crucible, lid, and contents) to determine the mass of the remaining "non-volatile" product. Record any pertinent data on your data sheet. Repeat this process for a total of three trials.

Safety Notes

Handle the crucibles with caution since a cold one looks the same as one hot enough to burn you. Be cautious using the Bunsen burner.

5.4 Waste Disposal

Waste for this lab may be dissolved in tap water and flushed down the sink.

5.5 Calculations

From your data you should be able to determine two important masses. First, the mass of the compound remaining after heating is the mass of anhydrous magnesium sulfate, that is, $MgSO_4$. Second, the mass change, between final and initial samples, equals the grams of water lost. Convert grams of $MgSO_4$ to moles of $MgSO_4$ and grams of H_2O to moles of H_2O using the appropriate molar masses (g/mol). *The nearest whole number ratio for moles of H_2O/moles of $MgSO_4$ corresponds to the water of hydration.*

EXPERIMENT 5: Water Loss by Dehydration

DATA SHEET

	Trial 1	Trial 2	Trial 3
Experimental:			
Before heating:			
Mass of Empty Crucible and Lid	_____	_____	_____
Mass of Crucible, Lid, and Epsom Salt Sample ($MgSO_4 \cdot XH_2O$)	_____	_____	_____
After heating:			
Mass of Crucible, Lid, and Dehydrated Salt ($MgSO_4$)	_____	_____	_____
Calculations:			
Mass of Hydrated Epsom Salt Sample ($MgSO_4 \cdot XH_2O$)	_____	_____	_____
Mass of Dehydrated Salt ($MgSO_4$)	_____	_____	_____
Moles of Dehydrated Salt ($MgSO_4$)	_____	_____	_____
Mass of H_2O Lost	_____	_____	_____
Moles of H_2O Lost	_____	_____	_____
Ratio (Moles H_2O/Moles $MgSO_4$) *(Recorded to 2 dec. places)*	_____	_____	_____
Avg. Ratio		_____	

Based upon your results, what is the molecular formula for Epsom salt? _____

EXPERIMENT 5: Water Loss by Dehydration

POSTLAB EXERCISE

Your lab instructor will provide values for the blanks below.

1. How many grams of water are in _____ grams of $CuSO_4 \cdot 5H_2O$?

2. If a _____ gram sample of a hypothetical hydrate $Tl_3As \cdot XH_2O$ loses _____ grams of H_2O, what is the value of X? (Hint: Molar mass of Tl_3As = 688.12 grams/mol.)

Unknown to you, the following events occur. Would the following increase, decrease, or not change your calculated value of X for $MgSO_4 \cdot XH_2O$ from the expected value? Explain each answer.

3. Some sample spattered out of the crucible while heating.

4. When obtained, the sample was contaminated with an inert solid which was unaffected by heating.

EXPERIMENT 6: Precipitation, Stoichiometry, and Synthesis of Alum

PRELAB EXERCISE

Terms:

    Precipitate –

    Alum –

    Filtration –

    Filtrate –

Safety Warnings:

EXPERIMENT 6: Precipitation, Stoichiometry, and Synthesis of Alum

PRELAB EXERCISE

Consider that 4.6295 g of aluminum are reacted with excess potassium hydroxide and sulfuric acid as described in the experimental section. After filtering, the mass of $KAl(SO_4)_2 \cdot 12H_2O$ is found to be 98.38 g.

1. How many moles of aluminum were obtained?

2. How many moles of $KAl(SO_4)_2 \cdot 12H_2O$ could be formed from the amount of aluminum given?

3. With the amount of aluminum given, what is the theoretical yield of $KAl(SO_4)_2 \cdot 12H_2O$ in grams?

4. What is the actual yield of $KAl(SO_4)_2 \cdot 12H_2O$ in grams?

5. What is the percent yield of $KAl(SO_4)_2 \cdot 12H_2O$?

PRECIPITATION, STOICHIOME-TRY, AND SYNTHESIS OF ALUM

Objectives

1. To perform redox and precipitation reactions.

2. To prepare an alum.

3. To gain a greater understanding of stoichiometry.

4. To practice vacuum filtration.

6.1 INTRODUCTION

Aluminum is used in the manufacture of automobiles, aircraft, and aluminum cans. This is because of the low density, high tensile strength, and resistance to corrosion shown by aluminum. Aluminum cans are recycled into other aluminum products (like window screens or outdoor furniture) or are used in the production of various aluminum compounds. Alum is the most common of these compounds.

Alum is used in the preparation of some food products, tanning of leather, and paper-making. An alum is a special type of ionic compound referred to as a double salt. That means it possesses two types of metal cations as well as the anion (sulfate in this case). One of the cations is in the $+1$ oxidation state (e.g., K^+, Na^+, or Ag^+) and the other is in the $+3$ oxidation state (e.g., Al^{3+}, Fe^{3+}, or Co^{3+}).

In this experiment you will be making the alum named potassium aluminum sulfate dodecahydrate. Its formula is $KAl(SO_4)_2 \cdot 12H_2O$. (Note: The prefix *dodeca-* means "twelve.") The potassium cation will come from potassium hydroxide. The aluminum cation will be derived from the oxidation of the elemental form of aluminum. The sulfate ion will come from sulfuric acid.

The pertinent reactions for this lab are:

1. Reaction of the aluminum with hot potassium hydroxide:
$$2Al(s) + 2KOH(aq) + 6H_2O(l) \rightarrow 2K^+(aq) + 2Al(OH)_4^-(aq) + 3H_2(g)$$

2. Conversion of the $Al(OH)_4^-(aq)$ to aluminum hydroxide:
$$Al(OH)_4^-(aq) + H^+(aq) \rightarrow Al(OH)_3(s) + H_2O(l)$$

3. Reaction of the aluminum hydroxide with excess sulfuric acid:
$$Al(OH)_3(s) + 3H^+(aq) \rightarrow Al^{3+}(aq) + 3H_2O(l)$$

4. Formation of the alum upon cooling:

$$K^+(aq) + 2SO_4^{2-}(aq) + Al^{3+}(aq) + 12H_2O(l) \rightarrow KAl(SO_4)_2 \cdot 12H_2O(s)$$

6.2 EXAMPLE CALCULATIONS

Consider that 15.00 g of aluminum are reacted with excess potassium hydroxide and sulfuric acid as described in the experimental section. After filtering, 222.54 g of $KAl(SO_4)_2 \cdot 12H_2O$ were obtained.

1. What is the molar mass of $KAl(SO_4)_2 \cdot 12H_2O$, the alum?

$$
\left.
\begin{array}{lll}
K: & 1 \times 39.10 \\
Al: & 1 \times 26.98 \\
S: & 2 \times 32.06 \\
O: & 20 \times 16.00 \\
H: & 24 \times 1.008
\end{array}
\right\}
\quad 474.39 \text{ g/mol}
$$

2. How many moles of aluminum were obtained?

$$15.00 \text{ g Al} \times \frac{\text{mol Al}}{26.98 \text{ g Al}} = 0.5560 \text{ mol Al}$$

3. How many moles of alum could be formed from the aluminum given?

$$0.5560 \text{ mol Al} \times \frac{2 \text{ Al(OH)}_4^-}{2 \text{ Al}} \times \frac{\text{Al(OH)}_3}{\text{Al(OH)}_4^-} \times \frac{\text{Al}^{3+}}{\text{Al(OH)}_3} \times \frac{\text{alum}}{\text{Al}^{3+}}$$

$$= 0.5560 \text{ mol alum}$$

4. With the aluminum given, what is the theoretical yield of $KAl(SO_4)_2 \cdot 12H_2O$?

$$0.5560 \text{ mol alum} \times \frac{474.39 \text{ g alum}}{\text{mol alum}} = 263.8 \text{ g alum}$$

5. What is the actual yield of alum?

$$222.54 \text{ g alum}$$

6. What is the percent yield of alum?

$$\frac{222.54 \text{ g alum}}{263.8 \text{ g alum}} \times 100\% = 84.37\%$$

6.3 EXPERIMENTAL

Note: As in any experiment, it is important to use clean glassware.

In a tared 250 mL beaker, weigh out between 1.00 and 1.20 g of solid aluminum. (*When given a range of measurements, aim for the minimum.*) Record the mass of aluminum sample obtained.

Slowly add 50 mL of 1.5 M KOH to the beaker. This will allow reaction 1 to take place. In this reaction, hydrogen gas is evolved, so you must keep your sample under the fume hood. Gently heat the beaker using a hot plate on a setting of 3 (or medium) to speed up the reaction. If froth develops, remove the beaker from the hotplate and place on wire gauze until the froth has dissipated. The froth will move aluminum particles up the side of the beaker; these particles need to be allowed to react with the KOH. Use a minimal amount of deionized water to wash these particles into the KOH solution and as needed to keep the liquid level in the beaker at around 60% of its original volume. Do not allow the liquid level to drop below half of its original volume. When hydrogen gas is no longer formed (the bubbling stops) the reaction is complete.

The resulting solution may not be clear. If it is not, vacuum filter the solution as demonstrated in Figure 1, and keep the filtrate (the liquid after a filtration step) for the next step. A second filtration may be necessary if the filtrate is still cloudy after the first filtration. Make sure to place the filtrate in a clean beaker after filtering.

Allow the solution to cool to room temperature. You may want to carefully run cool water over the outside of the beaker to aid the cooling.

Once cooled, slowly add 20 mL of 9 M H_2SO_4 while stirring the solution with a clean glass rod. A white precipitate of $Al(OH)_3$ will form as reaction 2 takes place. This solid will redissolve as more sulfuric acid is added and reaction 3 takes place.

Heat this solution over high heat to a low boil, stirring occasionally to dissolve the precipitate. Remove from the heat and allow the solution to cool for about 2 minutes. At this point, the solution should be clear. While the solution is cooling, prepare an ice bath by filling a large beaker (600–800 mL) halfway with crushed ice and adding enough tap water to barely cover the ice.

Carefully place the beaker into the ice bath and allow the solution to chill for about 20 minutes. If no crystals have formed after 10 minutes, take a clean glass stirring rod and gently rub the inside of the beaker. This helps speed the precipitation process. White crystals of alum should form during this time.

While the alum mixture is cooling, prepare another ice bath using a smaller beaker (250 mL recommended) to chill 50 mL of deionized water.

Prepare a filtration setup as shown in Figure 1.

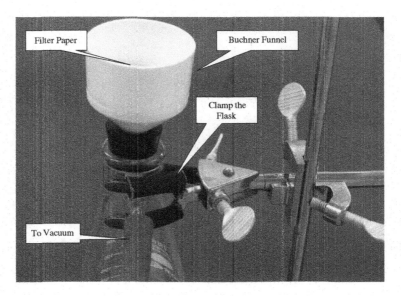

Figure 1 Vacuum Filtration

Filter the mixture using the aspirator vacuum (attached to the faucet) and Büchner funnel. Transfer as much of the solid to the Büchner funnel as possible. Use a clean glass stirring rod to help with the transfer.

Use about half of the chilled deionized water to rinse any remaining solid into the funnel and also to rinse the crystals. Repeat this step with the second half of the chilled water.

Continue the vacuum filtration for about 15 minutes to dry the crystals. The tip of a clean spatula may be used to spread out the crystals and speed the drying process.

Weigh a small clean beaker while the crystals are drying. Transfer all the crystals from the filter paper to the beaker. Avoid scraping paper particles into the beaker. Weigh the beaker with the crystals. Calculate and record the mass of alum formed.

Show your product to your lab instructor or helper and have him or her record any comments on its quality. Place the product in the labeled container. Clean and return the beaker to your drawer.

Calculate the theoretical yield given the starting mass of aluminum. Calculate the percent yield. Remember to show your work on a calculations page.

Safety Notes

KOH is a strong base. H_2SO_4 is a strong acid.

6.4 Waste Disposal

Place the alum in the provided collection container. Liquids may go down the drain.

EXPERIMENT 6: Precipitation, Stoichiometry, and Synthesis of Alum

DATA SHEET

Experimental:

Mass of Aluminum Obtained _____

Volume of 1.5 M KOH Used _____

Volume of 9 M H_2SO_4 Used _____

Mass of Empty Beaker _____

Mass of Beaker and Alum _____

Calculations:

Moles of Aluminum Obtained _____

Mass of Alum Formed _____

Theoretical Yield _____

Percent Yield _____

LA Notes concerning product quality (5 points):

EXPERIMENT 6: Precipitation, Stoichiometry, and Synthesis of Alum

POSTLAB EXERCISE

SHOW YOUR WORK!

1. Your percent yield should be less than 100%. Reread the procedure and give specific examples of what in this experiment could have prevented you from obtaining 100% of the theoretical yield.

Use the following information to answer the questions below:

_____ g of aluminum is/are reacted with excess KOH and H_2SO_4 as described in the experimental section. After filtering, the mass of $KAl(SO_4)_2 \cdot 12H_2O$ is found to be _____ g.

2. From the amount of aluminum given, what is the theoretical yield of $KAl(SO_4)_2 \cdot 12H_2O$ in grams?

3. What is the percent yield of $KAl(SO_4)_2 \cdot 12H_2O$?

EXPERIMENT 7: Solubility Study

PRELAB EXERCISE

Terms:

      Electrolyte –

      Cation –

      Anion –

      Precipitate –

      Spectator ion –

Safety Warnings:

EXPERIMENT 7: Solubility Study

PRELAB EXERCISE

To complete the following, consider the reaction between aqueous silver nitrate and aqueous potassium chloride. Be sure to balance the equation and to include the states of matter and the charges on ions.

1. Write the molecular equation.

2. Write the complete ionic equation.

3. Write the net ionic equation.

7

SOLUBILITY STUDY

Objectives

1. To use solubility guidelines to make predictions about the formation of precipitates from the mixing of various aqueous salt solutions.

2. To experimentally cause the formation of an ionic compound by mixing a cation with an anion.

3. To properly name an ionic compound.

4. To properly write the formula for an ionic compound.

5. To properly write molecular, complete ionic, and net ionic equations for precipitation reactions.

7.1 INTRODUCTION

When aqueous solutions of strong electrolytes are mixed, a chemical reaction may occur. With some of these reactions, there is no visible change, as in the case of an acid–base reaction. However, some reactions result in a visible change, such as the evolution of a gas or the formation of precipitate, an insoluble solid formed in solution from a chemical reaction between a cation and an anion. Whether or not a compound is soluble depends on several factors, but through experimental observations, a set of guidelines have been developed to help predict the solubility of many ionic compounds, commonly referred to as salts.

In this lab you will be using solubility guidelines to make predictions about the formation of precipitates when aqueous sodium (Na^+) solutions are mixed with aqueous nitrate (NO_3^-) solutions. To predict if a precipitate will form, consider the possible combinations of the ions present in the reactants. You will then test your predictions experimentally by mixing the equimolar solutions. Different backgrounds will be used to visually check for evidence of precipitate formation. It is possible, however, that minute amounts of precipitate, although present, may not be visible to the naked eye. Furthermore, because solubility is temperature and concentration-dependent, discrepancies between solubility guidelines and your experimental results may arise where solution temperatures and/or concentrations are different.

For the postlab, you will be asked to write the names for several cation–anion pairs and tentative formulas based on ion charges. (This may not represent the true identity of the compounds which are formed, but it is a good first guess based on your current knowledge of solution chemistry.) You will then write the molecular, complete ionic, and net ionic equations for a precipitation reaction.

Solubility Guidelines for Some Common Ionic Compounds in Water

Soluble Salts	Important Exceptions
Compounds containing	
Li^+, Na^+, K^+, NH_4^+	None
NO_3^-, $C_2H_3O_2^-$	None
Cl^-, Br^-, I^-	Compounds of Ag^+, Hg_2^{2+}, Pb^{2+}
SO_4^{2-}	Compounds of Sr^{2+}, Ba^{2+}, Ag^+, Pb^{2+}, Ca^{2+}

Insoluble (or slightly soluble) Salts	Important Exceptions
Compounds containing	
OH^-	Compounds of Li^+, Na^+, K^+, NH_4^+
	(Ca^{2+}, Sr^{2+}, and Ba^{2+} compounds are *slightly* soluble.)
S^{2-}	Compounds of Li^+, Na^+, K^+, NH_4^+, Ca^{2+}, Sr^{2+}, Ba^{2+}
CO_3^{2-} and PO_4^{3-}	Compounds of Li^+, Na^+, K^+, NH_4^+

Writing Molecular, Complete, and Net Ionic Equations

Several equations are used to represent a chemical reaction in solution. Molecular equations represent the reactants and products in undissociated form, whether they are strong or weak electrolytes, and show us the least about what actually takes place in a solution. In complete ionic equations all substances that are soluble strong electrolytes are represented as ions, including spectator ions, or ions that do not directly participate in the reaction. Spectator ions appear in identical forms on both the reactant and product sides of the complete ionic equation. Going a step further, net ionic equations eliminate spectator ions and any other component of a solution that is not directly involved in a reaction.

7.2 EXAMPLE CHEMICAL EQUATIONS

Molecular equation: reactants and products written in undissociated form

$$Pb(NO_3)_2\,(aq) + 2KI\,(aq) \rightarrow PbI_2\,(s) + 2KNO_3\,(aq)$$

Complete ionic equation: strong electrolytes represented as ions

* *PbI$_2$ is insoluble and therefore is **not** written as two separate ions.*

$$Pb^{2+}(aq) + 2NO_3^-(aq) + 2K^+(aq) + 2I^-(aq) \rightarrow {}^*PbI_2(s) + 2K^+(aq) + 2NO_3^-(aq)$$

Net ionic equation: spectator ions eliminated from the complete ionic equation

$$Pb^{2+}(aq) + 2I^-(aq) \rightarrow PbI_2(s)$$

7.3 EXPERIMENTAL

Before beginning the experiment, complete the table in Data Sheet 1 according to the directions given at the top of the sheet. The table shows the solutions you will be using in this experiment. *Why are sodium and nitrate solutions being used in this investigation?*

After you have made your predictions, obtain two well plates, each containing 24 small reaction wells. Thoroughly clean the plates and rinse with *deionized* water. Remove excess deionized water by vigorously shaking the well plates a few times. (Drying with a paper towel is not necessary and will leave lint in the wells.) Place the plates on your bench in horizontal orientation, one above the other. Together the plates will have six columns and eight rows. To ease identification, the columns are labeled with a numeral from 1 to 6, and the rows are labeled with a letter from A to D.

Starting in well 1A and going down the first column, place 3 drops of the NaCl solution into each of the eight wells. Continue in this manner with the remaining Na$^+$ solutions, starting at the top of a different column for each Na$^+$ solution and working in the order given in the table in Data Sheet 2.

Starting in well 1A and going across the first row, add 3 drops of NH$_4$NO$_3$ solution to each of the six wells and *gently* swirl to mix, being careful not to cross-contaminate solutions. Inspect each well for precipitation by observing it first on a black surface (e.g., the lab bench), then on a white surface (e.g., a plain piece of paper), and lastly over printed text (e.g., a page in the lab manual). If solid is visible in a well, if the mixture is hazy (Haziness may constitute a precipitate if it settles to the bottom.), or if the print appears indistinct when you look at it through the solution mixture, assume precipitation. Record **I** and the color of the precipitate in the table in Data Sheet 2. If no precipitate or haziness are visible, and the print is distinct when you look at it through the solution mixture, record **S** in the table. Continue in this manner with the remaining NO$_3^-$ solutions, starting with a different row for each NO$_3^-$ solution and working in the order given in the table in Data Sheet 2.

Cross-contamination of reagents could negatively affect your results. If you feel cross-contamination may have occurred during the procedure, or if you are unsure of a result, you should repeat the procedure for any mixture in question.

Safety Notes

At 0.2 M, the inorganic salt solutions used in today's experiment are fairly mild skin irritants. However, silver cation and lead are known toxins. Avoid contact with the eyes and skin. Wash your hands well before leaving the lab. Upon contact, $Fe(NO_3)_3$ stains skin and clothing.

7.4 WASTE DISPOSAL

In both academia and industry, the expense associated with hazardous chemical waste disposal is significant. Due to the hazardous nature of several of the ions in today's lab, waste from this experiment must be collected in an appropriate manner. To minimize the volume of waste produced, and thus the expense, first empty the well plate into a designated container. Then, using the wash bottle provided, rinse the plate with a *minimum* of water. Dispose of the rinsate in the same waste container.

EXPERIMENT 7: Solubility Study

DATA SHEET 1

SOLUBILITY PREDICTIONS

Before beginning the experiment, use the solubility guidelines given in the write-up to make predictions about whether or not a precipitate will form when the following aqueous salt solutions are mixed. Use the following abbreviations to indicate your predictions in the table below.

Abbreviations: **S** (soluble, no precipitation should occur)
 I (insoluble or slightly soluble, precipitation should occur)

	NaCl	NaBr	NaOH	Na_2CO_3	Na_2SO_4	Na_3PO_4
NH_4NO_3						
$Ca(NO_3)_2$						
$Ba(NO_3)_2$						
$Al(NO_3)_3$						
$Fe(NO_3)_3$						
$Cu(NO_3)_2$						
$AgNO_3$						
$Pb(NO_3)_2$						

EXPERIMENT 7: Solubility Study

DATA SHEET 2

DATA AND OBSERVATIONS:

Abbreviations: **S** (soluble, no precipitation occurred)
                I (insoluble or slightly soluble, precipitation occurred)

	NaCl	NaBr	NaOH	Na_2CO_3	Na_2SO_4	Na_3PO_4
NH_4NO_3						
$Ca(NO_3)_2$						
$Ba(NO_3)_2$						
$Al(NO_3)_3$						
$Fe(NO_3)_3$						
$Cu(NO_3)_2$						
$AgNO_3$						
$Pb(NO_3)_2$						

EXPERIMENT 7: Solubility Study

POSTLAB EXERCISE

1. List the discrepancies between your predictions and your experimental results and give possible explanations for them. Cross-contamination and human error are *not* valid explanations as you were instructed to repeat the procedure for any mixture producing questionable results. (5 points)

2. Record an appropriate name and chemical formula for each cation/anion pair below: (1/2 point per blank)

	Name	Formula
Cl^-		
Ag^+	_____	_____
NH_4^+	_____	_____
Ca^{2+}	_____	_____
Al^{3+}	_____	_____
Fe^{3+}	_____	_____
SO_4^{2-}		
Ag^+	_____	_____
NH_4^+	_____	_____
Ca^{2+}	_____	_____
Al^{3+}	_____	_____
Fe^{3+}	_____	_____

Instructor provided anion:

	Name	Formula
Ag^+	_____	_____
NH_4^+	_____	_____
Ca^{2+}	_____	_____
Al^{3+}	_____	_____
Fe^{3+}	_____	_____

3. Your lab instructor will provide you with a precipitation reaction. Write the molecular, complete ionic, and net ionic equations for this reaction. Be sure each equation is balanced.

EXPERIMENT 8: Introduction to Titrations: Preparation of a Standard Solution

PRELAB EXERCISE

Terms:

Titration –

Indicator –

Endpoint of a titration –

Equivalence point of a titration –

Neutralization reaction –

Primary standard –

Secondary standard –

Safety Warnings:

EXPERIMENT 8: Introduction to Titrations: Preparation of a Standard Solution

PRELAB EXERCISE

1. Potassium hydrogen phthalate $KHC_8H_4O_4$ (called KHP, for short) is a good choice for a weak acid primary standard. What is the molar mass of KHP?

2. To how many decimal places should any burette reading be made?

3. A solution is labeled as 6 M NaOH. How many milliliters of this solution would be needed to prepare 500 mL of an approximately 0.5 M NaOH solution?

4. In a titration, how close should the endpoint be to the equivalence point?

Use the following information to answer the questions below:

It is known that 2.62 g of KHP requires 17.38 mL of NaOH solution in an acid–base titration to reach the equivalence point. The molecular form of the reaction is:

$$HKC_8H_4O_4(aq) + NaOH(aq) \rightarrow NaKC_8H_4O_4(aq) + HOH(l)$$

5. How many moles of KHP are used during this titration?

6. How many moles of NaOH are used during this titration?

7. What is the concentration (i.e., molarity) of the NaOH solution?

INTRODUCTION TO TITRATIONS: PREPARATION OF A STANDARD SOLUTION

Objectives

1. To prepare and standardize a NaOH solution.

2. To learn proper titration technique.

8.1 INTRODUCTION: STANDARDIZING A SOLUTION

The reaction of an acid (H^+ donor) with a base (H^+ accepter) is one of the most common reactions in chemistry. An acid–base reaction is also called a neutralization reaction. This reaction can be described in general as:

$$Acid + Base \rightarrow Salt + Water$$

or

$$H_3O^+ (aq) + OH^- (aq) \rightarrow 2H_2O(l)$$

or

$$H^+(aq) + OH^-(aq) \rightarrow HOH(l)$$

Note: Water is written as HOH(l) in this experiment to make it easier to visualize reaction of the hydronium and hydroxide ions in an acid–base reaction.

The analytical technique where a reaction is carried out with the controlled addition of one reactant to another until both are consumed is called *titration*. An acid–base titration is the progressive addition of a base solution to an acid solution (or vice versa), typically using a burette. The concentration of the acid or base can then be determined from calculations applied to the titration data.

A standard is a solution whose composition and concentration is well known. *Primary standards* are made from compounds that have a relatively high molar mass and are readily available at relatively low cost and high purity. An example of a good primary standard is potassium hydrogen phthalate, $KHC_8H_4O_4$. It is commonly referred to as KHP. Its high molar mass (204.23 g/mol) as well as its availability at low cost and high purity make it an almost ideal primary standard. *Secondary standards* lack at least one of the criteria to make them primary standards. An example of a secondary standard is sodium hydroxide, NaOH. It has a relatively low molar mass (40.00 g/mol), which gives us one less significant figure for calculations. Pure NaOH also readily absorbs water, which prevents knowing exactly how much of it is weighed out. This makes NaOH a secondary standard. Secondary standard solutions are prepared to approximate concentrations and then standardized against a primary standard to determine their exact concentration.

In this experiment a NaOH solution is prepared and standardized using the primary standard potassium hydrogen phthalate, $KHC_8H_4O_4$. The titrant is the solution which is delivered in incremental amounts from the burette. The titrant may be an acid or a base. At one point in the addition, the number of moles of base in the titrant will completely react with the number of moles of acid in the sample (or vice versa). This point is called the *equivalence point of the titration*. Note: The equivalence point means that there are no remaining reactants, not that their numbers are equal; remember stoichiometry. The *endpoint of a titration* is when the titration is stopped. Visualization of the endpoint of the reaction occurs by an indicator. With good technique, the endpoint will be no more than one drop beyond the equivalence point.

The molecular form of the reaction taking place is:

$$HKC_8H_4O_4(aq) + NaOH(aq) \rightarrow NaKC_8H_4O_4(aq) + HOH(l)$$

An *indicator* is a dye that changes color depending on the composition of the solution into which it is placed. In this case, in the presence of a very small amount of excess base or acid, the indicator will change color. This point, which must be located as accurately as possible, signals the end of the titration. In other words, the correct volume of base (or acid) has been added from the burette. Visual indicators are frequently used to locate the endpoint in acid–base titrations.

Less than one drop of excess NaOH solution beyond the equivalence point will make the solution basic (containing excess OH^-). Phenolphthalein will be used as an indicator in the experiment. It changes from colorless to pink to indicate the presence of excess base so that the endpoint can be easily and accurately detected.

8.2 EXAMPLE CALCULATIONS

The following questions refer to the information below:

It is known that 6.21 g of KHP requires 35.23 mL of NaOH solution in an acid–base titration to reach the equivalence point. The molecular reaction is:

$$HKC_8H_4O_4(aq) + NaOH(aq) \rightarrow NaKC_8H_4O_4(aq) + HOH(l)$$

1. How many moles of KHP are used during this titration?

$$6.21 \ g \ KHP \times \frac{1 \ mol \ KHP}{204.22 \ g \ KHP} = 0.0304_{08383} \ mol \ KHP$$

2. How many moles of NaOH are used during this titration?

$$0.0304_{08383} \ mol \ KHP \times \frac{1 \ mol \ NaOH}{1 \ mol \ KHP} = 0.0304_{08383} \ mol \ NaOH$$

3. What is the concentration (i.e., molarity) of the NaOH solution?

$$\frac{0.0304_{08383} \ mol \ NaOH}{35.23 \ mL \ Solution} \times \frac{mL \ Solution}{10^{-3} \ L \ Solution} = \frac{0.863 \ mol \ NaOH}{L \ Solution} = 0.863 \ M \ NaOH$$

8.3 EXPERIMENTAL

Obtain further instruction on the use of a burette from your lab instructor. For this experiment you will be using a 50 mL burette. You will notice that the burette is graduated to the tenth of a milliliter. You will recall that readings should be taken one decimal beyond the graduation on the device. This means that the burette should be read to two decimal places. The burette should be fairly clean and unbroken. Note any chips or cracks, if present, and report them to your instructor.

8.4 PREPARATION AND STANDARDIZATION OF A NaOH SOLUTION

For this portion of the experiment, a concentrated solution of NaOH (normally between 3.0 and 6.0 M) will be provided by your instructor. Record its concentration, and calculate the volume of this solution, which must be diluted to 250 mL to provide a 0.2xx M NaOH solution. Obtain a 250 mL plastic bottle with cap. Rinse it with deionized water. It is not necessary to dry it. Using a clean graduated cylinder, obtain the calculated volume of the "concentrated" NaOH solution and transfer it to your plastic bottle. Add enough water to the bottle to make the total volume about 250 mL. It is not necessary to be exact since you will be determining the "exact" concentration of the NaOH by titration. Cap the bottle tightly and invert well to mix. Note: The numbers from this section are not used for further calculations.

Special Safety Note: Concentrated NaOH solutions are particularly harmful to your eyes. Rinse your eyes for 20–30 minutes if contact occurs. Keep your eye protection on!

Prepare your burette for use as described in "Appendix D: Burette". Fill the burette with the NaOH solution that you have prepared.

If they are not already out, obtain a magnetic stirrer and a magnetic stir bar. Clean the magnetic stir bar. Place the magnetic stirrer under the burette. Obtain a clean 250 mL Erlenmeyer flask.

Label the flask so that you can determine which trial you are working with. Place the flask on the balance. Tare the flask and weigh between 1.10 and 1.30 g of KHP into each flask. (*When given a range of measurements, aim for the minimum.*) Record the mass of KHP on your data sheet for the appropriate trial. Remove the flask from the balance. Dissolve[1] the sample in about 75 mL of deionized water, adding 1–2 drops of phenolphthalein indicator and a magnetic stir bar into the flask. To be sure all KHP is reacted, use a small amount of deionized water to wash down any KHP from the sides of the flask.

Place the Erlenmeyer flask on the magnetic stirrer. Turn on the magnetic stirrer so that the stir-bar is gently moving. **Do not** turn on the heating component of the magnetic stirrer. The magnetic stirrer will mix the components of the flask.

Turn the stop valve of the burette to allow the NaOH solution to slowly drain into the flask.

---

1 Note: At the start of the titration, quickly adding 8 to 10 mL of NaOH to your Erlenmeyer flask and swirling will dissolve any undissolved KHP and will not alter your final results!

Be careful to not drain the NaOH solution below the markings of the burette. If you get to the 45 mL mark on the burette without reaching the endpoint, consult your instructor. Your instructor will determine if you are making a mistake in your procedure. They may give you a procedure for how to refill the burette and still know the volume that you deliver to reach the endpoint.

You will notice a localized appearance of pink where the NaOH solution enters the flask. This is the NaOH interacting with the phenolphthalein indicator before finding the KHP. You will use this to help you accurately determine the endpoint. When the localized appearance of pink becomes more pronounced and persistent, reduce the flow of the NaOH solution to dropwise.

When 1 drop of the NaOH solution produces a faint pink throughout the entire liquid in the flask that persists for at least 30 seconds you are at the endpoint. Rinse the tip of the burette into the flask using deionized water. Record the final burette reading to the nearest 0.01 mL. Consult your instructor if you are having difficulty.

Repeat this process to have a total of four titrations.

When finished, clean up your work area and the balance area. **Return the magnetic stir bar to the side-shelf. Do not place them in the drawer of your work area.**

Safety Notes

NaOH is a strong base. (See previous warning.)

8.5 WASTE DISPOSAL

Waste for this lab may be flushed down the sink with tap water.

EXPERIMENT 8: Introduction to Titrations: Preparation of a Standard Solution

DATA SHEET

Experimental:

*Concentration of NaOH stock solution obtained (from label)   _____

*Volume of NaOH stock solution obtained   _____

	Trial 1	**Trial 2**	**Trial 3**	**Trial 4**
Mass of #KHP	_____	_____	_____	_____
Final Burette Reading	_____	_____	_____	_____
Initial Burette Reading	_____	_____	_____	_____

Calculations:

	Trial 1	**Trial 2**	**Trial 3**	**Trial 4**
Volume of NaOH Required	_____	_____	_____	_____
Moles of KHP	_____	_____	_____	_____
Moles of NaOH Required	_____	_____	_____	_____
Molarity of NaOH Solution	_____	_____	_____	_____

Avg. NaOH Concentration   _____

*Note: This is not used for further calculations.
#KHP is the abbreviation for potassium hydrogen phthalate. Its molar mass is 204.23 g/mol.

Note: Due to the stoichiometry of KHP and NaOH, the reaction is considered 1:1 so that moles of KHP = moles of NaOH required.

EXPERIMENT 8: Introduction to Titrations: Preparation of a Standard Solution

POSTLAB EXERCISE

SHOW YOUR WORK!

Use the following information to answer the questions below:

_____ g of KHP requires _____ mL of NaOH solution in an acid–base titration to reach the equivalence point. The molecular form of the reaction is:

$$HKC_8H_4O_4(aq) + NaOH(aq) \rightarrow NaKC_8H_4O_4(aq) + HOH(l)$$

1. How many moles of KHP are used during this titration?

2. How many moles of NaOH are used during this titration?

3. What is the concentration of the NaOH solution?

EXPERIMENT 9: Titration Analysis of Vitamin C

PRELAB EXERCISE

Terms:

Titration –

Titrant –

% (m/m) –

Vitamin C –

Basic –

Safety Warnings:

EXPERIMENT 9: Titration Analysis of Vitamin C

PRELAB EXERCISE

Use the following information to answer the questions below:

A NaOH solution was prepared as described in last week's lab and found to be 0.251 M. It required 21.82 mL of this NaOH solution to completely react with the ascorbic acid in a 1000 mg vitamin C tablet. The mass was determined to be 1103.06 mg. The molecular form of the reaction is

$$HC_6H_7O_6(aq) + NaOH(aq) \rightarrow NaC_6H_7O_6(aq) + H_2O(l)$$

Note: The tablet is not pure vitamin C.

1. How many moles of OH⁻ are used in the titration?

2. How many moles of vitamin C are in the tablet?

3. How many milligrams of vitamin C are in the tablet?

4. What is the percent by mass of vitamin C in the tablet?

TITRATION ANALYSIS
OF VITAMIN C

Objectives

1. To prepare and standardize a NaOH solution.

2. To learn proper titration technique.

3. To determine the number of milligrams of vitamin C in a vitamin C tablet.

4. To determine the percent by mass of vitamin C in a vitamin C tablet.

9.1 INTRODUCTION

Refer to the previous lab for information concerning preparing standard solutions for titrations.

In this experiment the amount of vitamin C in a tablet is determined by delivering a solution of base, the titrant, from a burette to the vitamin C in an Erlenmeyer flask. The *titrant*, therefore, is the solution that is delivered in incremental amounts from the burette. The titrant may be an acid or a base, but in this case it is the strong base sodium hydroxide. At one point in the addition, the number of moles of base in the titrant will completely react with the number of moles of acid in the sample (or vice versa). This point is called the equivalence point of the titration. The endpoint of a titration is when the titration is stopped. Visualization of the endpoint of the reaction occurs by an indicator.

An indicator is a dye that changes color depending on the composition of the solution into which it is placed. In this case in the presence of a very small amount of excess base or acid, the indicator will change color. This point, which must be located as accurately as possible, signals the end of the titration. In other words, the correct volume of base (or acid) has been added from the burette. Visual indicators are frequently used to locate the endpoint in acid–base titrations.

Ascorbic acid, $HC_6H_7O_6$, (common name: vitamin C) will be studied in this experiment. The vitamin C tablet was obtained from a local drugstore. It contains vitamin C and fillers. Vitamin C is needed for metabolic reactions in both plants and animals. Ascorbic acid is a soluble weak acid existing predominately as $HC_6H_7O_6(aq)$ molecules in solution. It is also light sensitive and slowly undergoes air oxidation in solution. For the purpose of this experiment, it will be considered to be a monoprotic acid, only being able to "lose" the first hydrogen listed in the formula. It will react rapidly and quantitatively with OH^- (provided by NaOH) by the following molecular reaction:

$$HC_6H_7O_6(aq) + NaOH(aq) \rightarrow NaC_6H_7O_6(aq) + H_2O(l)$$

Less than one drop of excess OH^- beyond the equivalence point will make the solution *basic* (containing excess OH^-). Phenolphthalein will be used as an indicator in the experiment. It changes from colorless to pink to indicate the presence of excess base so that the endpoint can be easily and accurately detected.

9.2 EXAMPLE CALCULATIONS

1. What is the molar mass of vitamin C, $HC_6H_7O_6$?

$$\left.\begin{array}{l} \text{C: } 6 \times 12.01 \\ \text{H: } 8 \times 1.008 \\ \text{O: } 6 \times 16.00 \end{array}\right\} \quad 176.12 \text{ g/mol}$$

The following questions refer to the information below:

A NaOH solution was prepared as described in last week's lab and found to be 0.0999 M. It required 39.78 mL of this NaOH solution to completely react with the vitamin C in a tablet whose mass is 1350 mg. The molecular form of the reaction is below:

$$HC_6H_7O_6(aq) + NaOH(aq) \rightarrow NaC_6H_7O_6(aq) + H_2O(l)$$

Note: The tablet is not pure vitamin C.

2. How many moles of OH^- are used in the titration?

NaOH is a strong base and will completely dissociate to give $Na^+(aq)$ and $OH^-(aq)$.

$$39.78 \text{ mL NaOH}_{Sln} \times \frac{10^{-3} \text{ L NaOH}_{Sln}}{\text{mL NaOH}_{Sln}} \times \frac{0.0999 \text{ mol NaOH}}{\text{L NaOH}_{Sln}}$$

$$= 3.97_{4022} \times 10^{-3} \text{ mol NaOH}$$

3. How many moles of vitamin C are in the tablet?

Use the balanced equation to go from what you are given information about to what you are asked to solve for.

$$3.97_{4022} \times 10^{-3} \text{ mol NaOH} \times \frac{1 \text{ mol HC}_6H_7O_6}{1 \text{ mol NaOH}} = 3.97_{4022} \times 10^{-3} \text{ mol HC}_6H_7O_6$$

4. How many milligrams of vitamin C are in the tablet?

$$3.97_{4022} \times 10^{-3} \text{ mol HC}_6H_7O_6 \times \frac{176.12 \text{ g HC}_6H_7O_6}{\text{mol HC}_6H_7O_6} \times$$

$$\frac{\text{mg HC}_6H_7O_6}{10^{-3} \text{g HC}_6H_7O_6} = 700 \text{ mg HC}_6H_7O_6$$

5.   What is the percent by mass of vitamin C in the tablet?

$$\% \ (m/m) = \frac{\text{mass of vitamin C in the tablet}}{\text{total mass of tablet}} \times 100\%$$

$$\frac{700 \text{ mg HC}_6\text{H}_7\text{O}_6}{1350 \text{ mg tablet}} \times 100\% = 51.9\%$$

9.3 EXPERIMENTAL

In this study, you will be performing two different sets of titrations. First you will prepare and standardize a NaOH solution. Then you will use the "standardized" NaOH to ultimately determine the number of milligrams of vitamin C in a tablet.

A NaOH solution will be prepared and standardized against the primary standard, potassium hydrogen phthalate, $KHC_8H_4O_4$. This standardization process is similar to the one you used last week.

Obtain further instruction on the use of a burette from your lab instructor. For this experiment you will be using a 50 mL burette. You will notice that the burette is graduated to the tenth of a milliliter. You will recall that readings should be taken 1 decimal beyond the graduation on the device. This means that the burette should be read to two decimal places. The burette should be fairly clean and unbroken. Note any chips or cracks, if present, and report them to your instructor.

Part A: Preparation and Standardization of a NaOH Solution

For this portion of the experiment, a concentrated solution of NaOH (normally between 3.0 and 6.0 M) will be provided by your instructor. Record its concentration, and calculate the volume of this solution, which must be diluted to 250 mL to provide a 0.2xx M NaOH solution. Obtain a 250 mL plastic bottle with cap. Rinse it with deionized water. It is not necessary to dry it. Using a clean graduated cylinder, obtain the calculated volume of the "concentrated" NaOH solution and transfer it to your plastic bottle. Add enough water to the bottle to make the total volume about 250 mL. It is not necessary to be exact since you will be determining the "exact" concentration of the NaOH by titration. Cap the bottle tightly and invert well to mix. Note: The numbers from this section are not used for further calculations.

Special Safety Note: Concentrated NaOH solutions are particularly harmful to your eyes. Rinse your eyes for 20 to 30 minutes if contact occurs. Keep your eye protection on!

Prepare your burette for use as described in "Appendix D: Burette". Fill the burette with the NaOH solution that you have prepared.

If they are not already out, obtain a magnetic stirrer and a magnetic stir bar. Clean the magnetic stir bar. Place the magnetic stirrer under the burette. Obtain a clean 250 mL Erlenmeyer flask.

Label the flask so that you can determine which trial you are working with. Place the flask on the balance. Tare the flask and weigh between 1.10 and 1.30 g of KHP into each flask. (*When given a range of measurements, aim for the minimum.*) Record the mass of KHP on your data sheet for the appropriate trial. Remove the flask from the balance. Dissolve* the sample in about 75 mL of deionized water, and add 1–2 drops of phenolphthalein indicator and a magnetic stir bar into the flask. To be sure all KHP is reacted, use a small amount of deionized water to wash down any KHP from the sides of the flask.

Place the Erlenmeyer flask on the magnetic stirrer. Turn on the magnetic stirrer so that the stir-bar is gently moving. **Do not** turn on the heating component of the magnetic stirrer. The magnetic stirrer will mix the components of the flask.

Turn the stop valve of the burette to allow the NaOH solution to slowly drain into the flask.

Be careful to not drain the NaOH solution below the markings of the burette. If you get to the 45 mL mark on the burette without reaching the endpoint, consult your instructor. Your instructor will determine if you are making a mistake in your procedure. They may give you a procedure for how to refill the burette and still know the volume that you deliver to reach the endpoint.

You will notice a localized appearance of pink where the NaOH solution enters the flask. This is the NaOH interacting with the phenolphthalein indicator before finding the KHP. You will use this to help you accurately determine the endpoint. Rinse the tip of the burette into the flask using deionized water. When the localized appearance of pink becomes more pronounced and persistent, reduce the flow of the NaOH solution to dropwise.

When 1 drop of the NaOH solution produces a faint pink throughout the entire liquid in the flask that persists for at least 30 seconds, you are at the endpoint. Record the final burette reading to the nearest 0.01 mL. Consult your instructor if you are having difficulty.

Repeat this process to have a total of three titrations.

*Note: At the start of the titration, quickly adding 8 to 10 mL of NaOH to your Erlenmeyer flask and swirling will dissolve any undissolved KHP and will not alter your final results!

Part B: Titration of Vitamin C Tablets

Obtain 3 vitamin C tablets and record the brand and stated potency for the tablets. **From this point on, you need to keep track of which tablet is which.** Determine the mass of each and record.

Place each in a separate clean 250 mL Erlenmeyer flask along with about 75 mL of deionized water. Allow the tablets to set in the water for about 5 minutes and then carefully break them up with the aid of a glass stirring rod. Some of the binder and inactive portions of the tablets will not dissolve, leaving a cloudy solution. Stir and swirl for 5–10 minutes to ensure the dissolution of the ascorbic acid. Rinse the stirring rod with a little deionized water to avoid loss of sample. If sample is stuck onto the stirring rod, leave it in the flask during the titration. Add 1–2 drops of phenolphthalein indicator.

Refill the burette from Part A with the now-standardized NaOH solution. Titrate the first vitamin C sample until you see a color change that persists for at least 30 seconds to signal the endpoint. Binders and dyes will make the endpoint more difficult to spot.

If the first titration of the vitamin C requires less than 20 mL of NaOH solution it is unnecessary to refill the burette as long as at least 20 mL of solution remains. This means that you can start your second (or third) titration from the previous endpoint reading if sufficient NaOH solution remains.

Calculate the number of milligrams of vitamin C in each tablet and the average amount of vitamin C per tablet in milligrams.

When finished, clean up your work area and the balance area. **Return the magnetic stir bars to the side-shelf. Do not place them in the drawer of your work area.**

Safety Notes

NaOH is a strong base. (See the previous Special Safety Note.)

9.4 WASTE DISPOSAL

Waste for this lab may be flushed down the sink with tap water.

EXPERIMENT 9: Titration Analysis of Vitamin C

DATA SHEET 1

A. Preparation and Standardization of a NaOH Solution

Experimental:

*Concentration of NaOH stock solution obtained (from label)   _____

*Volume of NaOH stock solution obtained   _____

	Trial 1	Trial 2	Trial 3
Mass of #KHP	_____	_____	_____
Final Burette Reading	_____	_____	_____
Initial Burette Reading	_____	_____	_____

Calculations:

	Trial 1	Trial 2	Trial 3
Volume of NaOH Required	_____	_____	_____
Moles of KHP	_____	_____	_____
Moles of NaOH Required	_____	_____	_____
Molarity of NaOH Solution	_____	_____	_____

Avg. NaOH Concentration   _____

*Note: This is not used for further calculations.
#KHP is the abbreviation for potassium hydrogen phthalate. Its molar mass is 204.23 g/mol.

Note: Due to the stoichiometry of KHP and NaOH, the reaction is considered 1:1 so that moles of KHP = moles of NaOH required.

EXPERIMENT 9: Titration Analysis of Vitamin C

DATA SHEET 2

B. Titration of Vitamin C Tablets

Experimental:

Calculated Concentration of NaOH (bottom of Data Sheet 1)   _____

Brand of Vitamin C Tablet   _____

Label Potency of Tablet   _____

	Trial 1	**Trial 2**	**Trial 3**
Mass of Tablet	_____	_____	_____
Final Burette Reading	_____	_____	_____
Initial Burette Reading	_____	_____	_____

Calculations:

	Trial 1	**Trial 2**	**Trial 3**
Volume NaOH Required	_____	_____	_____
Moles of NaOH Required	_____	_____	_____
Moles of Vitamin C	_____	_____	_____
Mass of Vitamin C (mg)	_____	_____	_____
% (m/m) Vitamin C in Tablet	_____	_____	_____

Avg. mg of Vitamin C   _____

Avg. % (m/m) of Vitamin C in the Tablet   _____

EXPERIMENT 9: Titration Analysis of Vitamin C

POSTLAB EXERCISE

SHOW YOUR WORK!

1. _____ g of KHP requires _____ mL of an approximately 0.3 M NaOH solution in an acid–base titration to reach the equivalence point. The molecular form of the reaction is

$$HKC_8H_4O_4(aq) + NaOH(aq) \rightarrow NaKC_8H_4O_4(aq) + H_2O(l)$$

What is the molarity of the NaOH solution?

2. _____ mL of the NaOH solution in question 1 are required to completely react with the ascorbic acid in a 1000 mg vitamin C tablet with a mass of _____ g. The molecular form of the reaction is

$$HC_6H_7O_6(aq) + NaOH(aq) \rightarrow NaC_6H_7O_6(aq) \rightarrow H_2O(l)$$

What is the mass percent of vitamin C in the tablet? Note: The tablet is not pure vitamin C.

EXPERIMENT 10: Ideal Gas Law: Estimation of Molar Mass of Magnesium

PRELAB EXERCISE

Terms:

      Molar mass –

      Ideal gas law –

Safety Warnings:

EXPERIMENT 10: Ideal Gas Law: Estimation of Molar Mass of Magnesium

PRELAB EXERCISE

A 0.0500 g sample of magnesium is reacted with 25 mL of 3.0 M sulfuric acid as described in the experimental section. The resulting hydrogen gas is collected over water and is found to have a volume of 51.0 mL. The temperature is 25 °C and the atmospheric pressure is 760.00 mm Hg. The column of water height inside the graduated cylinder is found to be 24.0 mm.

1. What is the height of the water column after converting to an equivalent height of mercury in mm?

2. What is the pressure of the hydrogen gas in the column in mm Hg?

3. What is the pressure in question 2 converted to atmospheres?

4. What is the number of moles of hydrogen gas in the sample?

5. What is the number of moles of metal in the sample?

6. What is the calculated molar mass of the metal for this trial?

7. What is the percent error in this determination?

10

IDEAL GAS LAW: ESTIMATION OF MOLAR MASS OF MAGNESIUM

Objectives

1. To perform a redox reaction involving the evolution of a gas.

2. To use stoichiometry.

3. To use the ideal gas law.

10.1 INTRODUCTION

The ideal gas law expresses the relationship between the number of moles (n), the pressure (P), the volume (V), and the temperature (T) of a gas. The equation for this relationship is:

$$PV = nRT$$

in which R is known as the ideal gas constant and has a value of 0.0821 L•atm/(mol•K).

The ideal gas law can be applied in determining the number of moles of gas present in a sample. If the gas is produced as a result of a redox reaction with magnesium and the net ionic equation is known, then the number of moles of magnesium in a sample can be determined. The net ionic equation is shown:

$$Mg(s) + 2H^+(aq) \rightarrow Mg^{2+}(aq) + H_2(g)$$

Therefore $n_{hydrogen} = n_{metal}$. If the metal is pure and its mass was determined before the reaction, then the molar mass of the metal is easy to calculate. The calculations are explained in more detail in the example calculations section.

10.2 EXAMPLE CALCULATIONS

A 0.0511 g sample of magnesium is reacted with 25 mL of 3.0 M sulfuric acid as described in the experimental section. The resulting hydrogen gas is collected over water and is found to have a volume of 52.7 mL. The temperature is 25 °C and the atmospheric pressure is 765.44 mm Hg. The column of water height inside the graduated cylinder is found to be 25.6 mm.

1. What is the height of the water column after converting to an equivalent height of mercury in mm?

    This conversion accounts for the differing densities of water and mercury. Mercury is 13.6 times more dense.

    $$25.6 \text{ mm } H_2O \times (1.00 \text{ mm } Hg/13.6 \text{ mm } H_2O) = 1.88 \text{ mm } Hg$$

2. What is the pressure of the hydrogen gas in the column in mm Hg?

    This is the atmospheric pressure minus the pressure of the water vapor, also accounting for the pressure difference inside the graduated cylinder and the outside atmosphere (the corrected height of the water column).

    $$765.44 - 23.8 - 1.88 = 739.7_6 \text{ mm } Hg$$

3. What is the pressure in question 2 converted to atmospheres?

    $$739.76 \text{ mm } Hg \times (1.00 \text{ atm}/760 \text{ mm } Hg) = 0.973_4 \text{ atm}$$

4. What is the number of moles of hydrogen gas in the sample?

    We will solve the ideal gas law for moles.

    $$n = \frac{PV}{RT} = \frac{0.973_4 \text{ atm} \times 0.0527 \text{L}}{0.0821 \text{ L} \cdot \text{atm}/(\text{mol} \cdot \text{K}) \times 298 \text{ K}} = 0.00209_7 \text{ mol } H_2$$

5. What is the number of moles of metal in the sample?

    We will use the balanced equation.

    $$0.00209_7 \text{ mol } H_2 \times \frac{1 \text{ mol Mg}}{1 \text{ mol } H_2} = 0.00209_7 \text{ mol Mg}$$

6. What is the calculated molar mass of the metal for this trial?

    $$\frac{0.0511 \text{ g Mg}}{0.00209_7 \text{ mol Mg}} = 24.37 \text{ g/mol Mg}$$

7. What is the percent error in this determination?

    $$\frac{|24.37 - 24.31|}{24.31} \times 100\% = 0.25\%$$

10.3 EXPERIMENTAL

Each student should obtain three strips of metal, each approximately 3 cm long, a 100 mL graduated cylinder, a large beaker (at least 800 mL), a thermometer, a ruler, a number 5 rubber stopper with a hole, a 2 inch wooden splint with a 1/2 cm slit in one end, and a ring stand with a three-prong clamp. Set up as demonstrated by your instructor.

Set up the ring stand and three-prong clamp to hold the graduated cylinder in order to prevent it from tipping over during the experiment. Fill the beaker half-full with tap water.

Weigh a metal strip to the nearest 0.0001 g and record this weight on your data sheet. Place the wooden splint in the stopper so that it does not stick out the top and at least 1.5 inches hangs below the stopper on the narrow side. Stretch plastic wrap over the top of the stopper to cover the hole. Insert one end of the metal strip into the slit in the splint. Wrap the metal strip around the wooden splint.

Pour 25 mL of 3.0 M sulfuric acid into the 100 mL graduated cylinder. Slowly fill the rest of the graduated cylinder with deionized water to the very top. It is important to **pour slowly** to prevent the water from mixing with the acid. Gently cap the graduated cylinder with the rubber stopper. Hold your finger over the lip of the graduated cylinder; invert the cylinder and **immediately** place it in the large beaker containing the water. The cylinder should not be on the bottom of the beaker, as the figure on the next page seems to indicate. At this point, clamp the cylinder to the ring stand. Place the thermometer in the beaker.

Note: The bubbles indicate hydrogen gas is being collected above the liquid. Once the metal dissolves and stops emitting bubbles, record the temperature of the water, which equals the temperature of the gas. Read the volume of the hydrogen gas from the graduated cylinder; you will have to read the graduated cylinder upside down. Record the graduated cylinder volume on the data sheet. Measure the difference in the water levels in the beaker and the graduated cylinder with a ruler and record on the data sheet.

Repeat this process for a total of three trials.

Use a barometer to read atmospheric pressure; remember to record this number on the data sheet. The atmospheric pressure is the total pressure of the gas in the cylinder plus the height of the water column. Since the gas was collected above water, the vapor pressure of water must be included in the calculation (Dalton's law of partial pressures). This value can be obtained for various temperatures from the table on the next page.

Safety Notes

Sulfuric acid is both a corrosive and an irritant. Wash off after contact. Do not force the stopper into the graduated cylinder as this will cause the cylinder to break.

10.4 WASTE DISPOSAL

All liquid waste can be poured down the sink with tap water. Any unreacted ribbon should be returned to the solid waste container provided for this experiment.

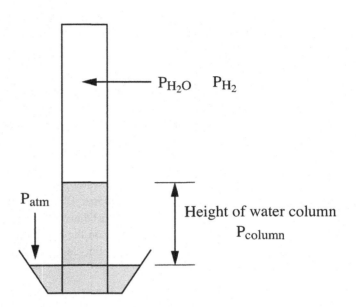

Vapor Pressures of Water

Temp (°C)	P_{H_2O} mm Hg
15	12.8
16	13.6
17	14.5
18	15.5
19	16.5
20	17.5
21	18.6
22	19.8
23	21.1
24	22.4
25	23.8
26	25.2
27	26.7
28	28.3
29	30.0

P_{col} must be corrected from mm H_2O to mm Hg by dividing by 13.6 (the density of Hg).

EXPERIMENT 10: Ideal Gas Law: Estimation of Molar Mass of Magnesium

DATA SHEET

	Trial 1	Trial 2	Trial 3
Experimental:			
Mass of Magnesium	_____	_____	_____
Volume of Hydrogen Gas (L)	_____	_____	_____
Temperature of Gas (K)	_____	_____	_____
Atmospheric Pressure (mm Hg)	_____	_____	_____
Vapor Pressure of Water According to Temperature (mm Hg)	_____	_____	_____
Height of Water Column (mm) (difference in water levels)	_____	_____	_____
Calculations:			
Mercury Equivalent Height (above line/13.6) = P_{col} (mm Hg)	_____	_____	_____
Pressure of Hydrogen (mm Hg)	_____	_____	_____
Pressure of Hydrogen (atm)	_____	_____	_____
Moles of Hydrogen Gas	_____	_____	_____
Moles of Magnesium	_____	_____	_____
Calculated Molar Mass of Magnesium (g/mol Mg)	_____	_____	_____
Avg. Molar Mass		_____	
% Error		_____	

EXPERIMENT 10: Ideal Gas Law: Estimation of Molar Mass of Magnesium

POSTLAB EXERCISE

A 0.0524 g sample of magnesium is reacted with 25 mL of 3.0 M sulfuric acid as described in the experimental section. The resulting hydrogen gas is collected over water and is found to have a volume of 53.6 mL. The temperature is 25 °C and the atmospheric pressure is 762.12 mm Hg. The column of water height inside the graduated cylinder is found to be 22.9 mm.

1. What is the height of the water column after converting to an equivalent height of mercury in mm?

2. What is the pressure of the hydrogen gas in the column in mm Hg?

3. What is the pressure in question 2 in atmospheres?

4. What is the number of moles of hydrogen gas in the sample?

5. What is the number of moles of metal in the sample?

6. What is the calculated molar mass of the metal for this trial?

7. What is the percent error in this determination?

EXPERIMENT 11: Specific Heat Capacity of Metals: Law of Dulong and Petit

PRELAB EXERCISE

Terms:

Calorimetry –

Specific heat capacity –

Molar heat capacity –

Dulong and Petit's law –

Safety Warnings:

EXPERIMENT 11: Specific Heat Capacity of Metals: Law of Dulong and Petit

PRELAB EXERCISE

1. If 26.03 g of water showed a temperature increase of 1.02 °C upon addition of hot metal, what is the heat change for the water?

2. Did the water absorb or give off heat?

3. What is the heat change by the metal in question 1?

4. Did the metal absorb or give off heat?

5. If the hot metal in question 1 has a mass of 5.43 g, and the metal underwent a temperature decrease of 14.45 °C, what is the specific heat capacity of the metal object?

SPECIFIC HEAT CAPACITY OF METALS: LAW OF DULONG AND PETIT

Objectives

1. To gain experience with coffee-cup calorimetry.

2. To determine the specific heat capacity and the molar heat capacity of a sample.

3. To gain an appreciation of the law of Dulong and Petit.

11.1 INTRODUCTION

Calorimetry is the measure of heat flow into or out of a system. The heat flow is measured in a device called a calorimeter. An ideal calorimeter would insulate the substance in the calorimeter so well that NO HEAT would be lost to the surroundings. The calorimeter you will use in this experiment (two nested Styrofoam coffee cups) is far from an ideal one, but we will "assume" that no heat flows in or out of the calorimeter.

When heat flows into or out of a substance the temperature of the substance usually changes. This change can be used to monitor the flow of heat energy. In order to determine the exact amount of heat (q) that flows, we need to know the temperature change in the substance (ΔT), the mass of the substance (m), and the specific heat capacity (*C*) of the substance.

Specific heat capacity is defined as the amount of heat (in joules) required to raise the temperature of 1 gram of the substance by 1 °C. The mathematical relationship relating the three quantities above is:

$$q = m\,C\Delta T$$

In this experiment you will measure the specific heat capacity of several metals by placing the hot metal in cold water. Heat will flow from the metal to the water until they reach the same final temperature.

The amount of heat absorbed by the water, q_{water}, can be calculated using the equation of

$$q_{water} = m_{water}\,C_{water}\Delta T_{water}$$

Since energy is conserved, the amount of heat absorbed by the water in the calorimeter is the amount of heat given off by the preheated metal. The heat change for the two processes is the same, but with opposite signs.

$$q_{metal} = {}^-q_{water}$$

The amount of heat the room temperature metal absorbed from the boiling water bath can be used to calculate C_{metal} using

$$q_{metal} = m_{metal} \, C_{metal} \Delta T_{metal}$$

(Note: The negative sign in the equation accounts for the fact that metal decreases in temperature and the water increases in temperature; i.e., an exchange of heat.)

The value of C_{water} = 4.184 J/(g•°C). After performing the experiment, the only unknown will be C_{metal}. This will be calculated from the other experimental data.

You will also experimentally validate **Dulong and Petit's law**, which says that metals will have a molar heat capacity of about 24.9 J/(mol•°C) (molar heat capacity = specific heat capacity × molar mass).

11.2 EXAMPLE CALCULATIONS

The following questions refer to the information below:

When 45.92 g of aluminum at 101.4 °C are added to 64.32 g of water at 24.8 °C, the resulting temperature of the water and metal is 35.1 °C.

1. What is the temperature change of the water?

$$35.1 \text{ °C} - 24.8 \text{ °C} = 10.3 \text{ °C}$$

2. How much heat did the water absorb?
    Note: the underlined digit in the answer is the last significant digit. Extra digits are kept until all calculations are completed.

$$64.32 \text{ g H}_2\text{O} \times 4.184 \, \frac{\text{J}}{\text{g} \cdot \text{°C}} \times 10.3 \text{ °C} = 27\underline{7}1.88 \text{ J}$$

3. What is the temperature change of the metal?
    Note: the temperature is negative because the metal loses heat.

$$35.1 \text{ °C} - 101.4 \text{ °C} = (-)66.3 \text{ °C}$$

4. How much heat did the metal give off?

$$q_{metal} = (-) \, q_{water} = (-)27\underline{7}1.88 \text{ J}$$

5. What is the specific heat capacity for the metal?

$$\frac{(-)27\underline{7}1.883 \text{ J}}{(45.92 \text{ g})(-66.3 \text{ °C})} = 0.91\underline{0}457 \, \frac{\text{J}}{\text{g} \cdot \text{°C}}$$

6. What is the molar heat capacity for the metal?

$$0.91\underline{0}457 \, \frac{\text{J}}{\text{g} \cdot \text{°C}} \times 26.98 \, \frac{\text{g Al}}{\text{mol Al}} = 24.\underline{5}641 \, \frac{\text{J}}{\text{mol} \cdot \text{°C}}$$

7. What is the percent error for the molar heat capacity of the metal?
    Note: the underlined digit in 0.3359 is the only significant digit because it was calculated using subtraction. Therefore, the answer has only 1 significant digit.

$$\frac{24.9 - 24.5641}{24.9} \times 100\% = \frac{0.\underline{3}359}{24.9} \times 100\% = \underline{1}.349\% \text{ or } 1\%$$

11.3 Experimental (Work in Pairs)

Set up a hot water bath and "coffee-cup calorimeter" as described by your instructor. Deionized water should be used in the calorimeter, but use tap water for the hot water bath.

Weigh between 60 and 80 g of your metal into a dry 100 mL beaker. It is important to have at least 60 g, but *if necessary*, your sample can slightly exceed 80 g. Carefully transfer the metal into a large dry test tube. Place this test tube into your boiling water bath, making sure that the metal in the tube is beneath the water level in the water bath. (*Note: Take care not to splash any of the water into the test tube; the metal should remain absolutely dry.*) Allow the metal to heat for about 15 to 20 minutes, thereby reaching the temperature of the boiling water (100 °C). Measure and record the actual temperature with your thermometer to the nearest 0.1 °C. This will be the initial temperature of the metal.

thermometer   stirrer

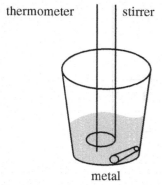

metal

Figure 1 Coffee-cup Calorimeter with Metal

While your metal is heating, dry your coffee cup calorimeter with a paper towel. Then determine the mass of about 40 mL of deionized water by weighing it directly in the tared calorimeter. Record the mass.

Carefully assemble your calorimeter and read the temperature of your water to the nearest 0.1 °C. This is the initial temperature of your water.*

Now, remove the test tube from the boiling water, wipe off the tube, and quickly empty the metal into your calorimeter without undue splashing. Cover the calorimeter and stir the contents. Monitor the temperature of the contents and record the HIGHEST temperature reached. This is the final temperature of the metal and the water.

Now carefully decant the water off of the metal, collect your metal in a paper towel and pat it as dry as possible. After drying the metal, perform the experiment above for a second, and then a third trial.

Calculate the average specific heat capacity for your metal. Record the name of your sample and your calculated SHC value in the Data Sheet, and then copy down the sample names and SHC values for two different metals tested by other students. Finally, calculate the molar heat capacity (MHC) for the three metals. How close is your MHC to Dulong and Petit's 24.9 J/(mol•°C)? Calculate the percent error of your value when compared to Dulong and Petit's value.

$$\% \ error = \frac{24.9 - determined \ value}{24.9} \times 100\%$$

Note: After measuring the boiling water temperature, dry your thermometer and allow it to cool to room temperature before placing it in your calorimeter! Cooling the thermometer can be accomplished by placing the thermometer under cool running water. Again, be certain the thermometer is dry before measuring the temperature of water in the calorimeter.

Safety Notes

Be cautious using the boiling water bath.

11.4 Waste Disposal

Return the metals to their original location. Pour the water down the drain.

NAME: _____   LAB SECTION: _____

PARTNER'S NAME: _____

EXPERIMENT 11: Specific Heat Capacity of Metals: Law of Dulong and Petit

DATA SHEET

	Trial 1	Trial 2	Trial 3
Experimental:			
Mass of Metal Sample	_____	_____	_____
Mass of Water in Calorimeter	_____	_____	_____
$T_{initial}$ of Metal (temperature of boiling water)	_____	_____	_____
$T_{initial}$ of Water (in calorimeter)	_____	_____	_____
T_{final} of Metal and Water	_____	_____	_____
Calculations:			
ΔT_{water}	_____	_____	_____
ΔT_{metal}	_____	_____	_____
q_{Water}	_____	_____	_____
q_{Metal}	_____	_____	_____
Specific Heat Capacity of Metal	_____	_____	_____
Avg. Specific Heat Capacity	_____		

	Metal	**Molar Mass**	**SHC***	**MHC****
1.	_____	_____	_____	_____
2.	_____	_____	_____	_____
3.	_____	_____	_____	_____

Avg. MHC _____

% Error _____

*SHC = Specific Heat Capacity   **MHC = Molar Heat Capacity

EXPERIMENT 11: Specific Heat Capacity of Metals: Law of Dulong and Petit

POSTLAB EXERCISE

1. If _____ g of water at 22.9 °C increased to _____ °C upon addition of 15.62 g of metal at 50.32 °C, how much heat did the water absorb?

2. What is the specific heat capacity of the metal in question 1?

3. If equivalent amounts of heat are absorbed by 1 mole of Metal A (molar mass = 78 g/mol) and 1 mole of Metal B (molar mass = 100 g/mol), which shows the greatest rise in temperature? (Hint: Assume the metals obey Dulong and Petit's law.) Explain your reasoning.

EXPERIMENT 12: Hess's Law

PRELAB EXERCISE

Terms:

Hess's law –

First law of thermodynamics –

Calorimetry –

Neutralization reaction –

Heat of neutralization –

Heat of solution –

ΔH^{o}_{f} –

Safety Warnings:

EXPERIMENT 12: Hess's Law

PRELAB EXERCISE

1. The heat of neutralization of HCl*(aq)* by NaOH*(aq)* is −55.84 kJ/mol H_2O produced. Fifty (50.00) mL of 1.73 M NaOH are added to 35.00 mL of 2.08 M HCl. Both solutions are at 22.80 °C. How many moles of water are produced by this reaction?

2. What is the final solution temperature of the solution in question 1, assuming that the final solution has a density of 1.02 g/mL and a specific heat of 3.98 J/(g•°C)?

Use the following information to answer the questions below:

$$2ZnS(s) + O_2(g) \rightarrow 2ZnO(s) + 2S(s) \qquad \Delta H_{rxn} = -290.8 \text{ kJ}$$

3. Calculate ΔH_{rxn} for

$$ZnS(s) + \tfrac{1}{2}O_2(g) \rightarrow ZnO(s) + S(s)$$

4. Calculate ΔH_{rxn} for

$$2ZnO(s) + 2S(s) \rightarrow 2ZnS(s) + O_2(g)$$

5. Given

$$Cu(s) + Cl_2(g) \rightarrow CuCl_2(g) \qquad\qquad \Delta H_{rxn} = -206 \text{ kJ}$$

$$2Cu(s) + Cl_2(g) \rightarrow 2CuCl(s) \qquad\qquad \Delta H_{rxn} = -136 \text{ kJ}$$

calculate the enthalpy change in the following reaction:

$$CuCl_2(g) + Cu(s) \rightarrow 2CuCl(s)$$

12

HESS'S LAW

Objectives

1. To gain experience with coffee-cup calorimetry.

2. To gain experience using Hess's law.

3. To gain experience with acid–base and solubility reactions.

12.1 INTRODUCTION

Hess's law states, as a direct consequence of the first law of thermodynamics, *that the energy change in a reaction is the same regardless of whether it occurs in a single step or in a series of stepwise chemical reactions.* The use of Hess's law is most important since in many cases thermochemical quantities cannot be determined experimentally by direct calorimetric methods. This is particularly true for standard enthalpies of formation, ΔH_f°, because direct synthesis of a compound from the constituent elements may not occur or may be a reaction that occurs too slowly to make calorimetric measurement feasible.

In this experiment you will be given values from the literature for the enthalpy of formation (298 K) of aqueous ammonia and aqueous HCl:

$$\tfrac{1}{2} N_2(g) + \tfrac{3}{2} H_2(g) \rightarrow NH_3(aq) \qquad \Delta H_f^\circ = -80.0 \text{ kJ} \qquad \text{(Equation 1)}$$

$$\tfrac{1}{2} H_2(g) + \tfrac{1}{2} Cl_2(g) \rightarrow HCl(aq) \qquad \Delta H_f^\circ = -165.0 \text{ kJ} \qquad \text{(Equation 2)}$$

The NH_3 and the HCl will be mixed in equimolar ratios so as to form quantitatively an aqueous solution of an ammonium salt, $NH_4Cl(aq)$; the temperature rise given by $\Delta T = T_{final} - T_{initial}$ will be measured for the "neutralization" reaction and used to calculate ΔH_{neut}:

$$NH_3(aq) + HCl(aq) \rightarrow NH_4Cl(aq) \qquad \Delta H_{neut} \qquad \text{(Equation 3)}$$

Likewise, ΔT will be measured for the dissolution of the solid ammonium salt, $NH_4Cl(s)$, in water:

$$NH_4Cl(s) \rightarrow NH_4Cl(aq) \qquad \Delta H_{soln} \qquad \text{(Equation 4)}$$

Finally, the appropriate two literature enthalpies will be combined with the experimental enthalpies, ΔH_{neut} and ΔH_{soln}, to give by Hess's law a value for the molar enthalpy of formation of the solid ammonium salt, that is, ΔH_f° for $NH_4Cl(s)$. The overall reaction is:

$$\tfrac{1}{2} N_2(g) + 2H_2(g) + \tfrac{1}{2} Cl_2(g) \rightarrow NH_4Cl(s) \qquad \text{(Equation 5)}$$

For reactions like these, occurring in situ, the heat absorbed by the surroundings (solution being heated) can be assumed to be equal to the heat evolved by the reaction (no heat loss to the container or outside the container). Therefore,

$$q_{rxn} = -q_{surr} = -m_{surr} \, C_{surr} \, \Delta T_{surr}$$

and

$$\Delta H_{rxn} = q_{rxn} \, / \text{ moles product}$$

where m, C, and ΔT refer to the mass (g), specific heat (J/(g•°C)), and $(T_{final} - T_{initial})$ for the surroundings, respectively. Note that the temperature measurement is made in the immediate thermal surroundings, the solution in which the reaction occurs. (The surroundings solution is aqueous and dilute (largely water), $d_{H_2O} = 1.00$ g/mL and $C_{H_2O} = 4.184$ J/(g•°C))

Therefore, q_{rxn} may be calculated in joules. Conversion to kilojoules and dividing by the number of moles of reactant or product gives the molar enthalpy change, ΔH_{rxn}, in kJ/mole. The final calculated results for the ΔH_f° for the $NH_4Cl(s)$ can be compared to that tabulated by the National Institute of Standards and Technology (NIST):

Solid Ammonium Salts	ΔH_f° (kJ/mole)
NH_4Cl	−315.4
NH_4Br	−270.3
NH_4NO_3	−365.1
$NH_4H_2PO_4$	−1,450.8

12.2 EXAMPLE CALCULATIONS

When 49.9 mL of 2.0 M NH_3 at 22.5 °C are added to 50.2 mL of 1.8 M HCl at 21.8 °C, the resulting temperature of the mixture is 30.6 °C. The mass of the mixture is found to be 98.74 g. The molecular reaction for the neutralization is:

$$NH_3(aq) + HCl(aq) \rightarrow NH_4Cl(aq)$$

1. What is the average initial temperature?

$$\frac{(22.5 \, °C \times 49.9 \, mL) + (21.8 \, °C \times 50.2 \, mL)}{(49.9 \, mL + 50.2 \, mL)} = 22.1\underline{4}90 \, °C$$

2. What is the temperature change of the water?

$$30.6 \, °C - 22.1\underline{4}90 \, °C = 8.\underline{4}51 \, °C$$

3. What is the heat flow in kJ for the reaction?

$$q_{mixture} : \quad \frac{98.74 \, g \times 4.184 \frac{J}{g \cdot °C} \times 8.\underline{4}51 \, °C}{1000 \frac{J}{kJ}} = 3.\underline{4}91 \, kJ$$

$$q_{neut} = (-)q_{mixture} = (-)3.\underline{4}91 \, kJ$$

4. How many moles of NH_4Cl are produced?

$$\left(\frac{2.0 \text{ mol}}{L}\right)\left(\frac{0.0499L}{}\right) = 0.09\underline{9}80 \text{ mol } NH_3$$

$$\left(\frac{1.8 \text{ mol}}{\text{l.}}\right)\left(\frac{0.0502\text{L}}{}\right) = 0.09\underline{0}36 \text{ mol HCl} \quad \textit{limiting reagent}$$

$$\left(\frac{0.09\underline{0}36 \text{ mol HCl}}{}\right)\left(\frac{1 \text{ mol NH}_4\text{Cl}}{1 \text{ mol HCl}}\right) = 0.09\underline{0}36 \text{ mol NH}_4\text{Cl}$$

5. What is the enthalpy change for the reaction?

$$\frac{(-)3.\underline{4}91 \text{ kJ}}{0.09\underline{0}36 \text{ mol NH}_4\text{Cl}} = (-)39 \text{ kJ/mol}$$

The following questions refer to the information below:

A 4.05 g sample of $NH_4Cl(s)$ is added to 99.3 mL of water at 24.1 °C. The lowest tempera-ture the mixture reaches is 19.4 °C, and the mass of the mixture is found to be 102.03 g. The molecular reaction for the dissolution of solid ammonium chloride in water is:

$$NH_4Cl(s) \rightarrow NH_4Cl(aq)$$

6. What is the temperature change of the water?

$$19.4 \text{ °C} - 24.1 \text{ °C} = (-)4.7 \text{ °C}$$

7. What is the heat flow in kJ for the reaction?

$$q_{mixture} = \frac{102.03 \text{ g} \times 4.184 \frac{J}{g \cdot °C} \times (-)4.7 \text{ °C}}{1000\frac{J}{kJ}} = (-)2.\underline{0}06 \text{ kJ}$$

$$q_{soln} = (-)q_{mixture} = 2.006 \text{ kJ}$$

8. How many moles of NH_4Cl are produced?

$$\frac{4.05 \text{ g}}{53.49 \frac{g}{mol}} = 0.075\underline{7}15 \text{ mol}$$

9. What is the enthalpy change for the reaction?

$$\frac{2.\underline{0}06 \text{ kJ}}{0.075\underline{7}15 \text{ mol}} = 26 \text{ kJ/mol}$$

12.3 EXPERIMENTAL

Obtain two Styrofoam cups and a lid to use as an insulated container (calorimeter) and a thermometer for carrying out the reactions.

Part A: Heat of Neutralization

Carefully dry, weigh, and record the mass of your calorimeter (cups, lid, and thermom-eter). Now obtain 50 mL of NH_3 in your 100 mL graduated cylinder and place it in the calorimeter. Obtain 50 mL of 1.5 M HCl in a second 100 mL graduated cylinder. Allow the two solutions to stand until their temperatures are the same (within ± 0.5 degrees). If they are not, you may obtain an average temperature by the formula:

$$(T_1V_1 + T_2V_2)/(V_1 + V_2)$$

where T and V represent <u>temperature</u> and <u>volume</u>, respectively. Be sure to rinse off and dry the thermometer when transferring between the solutions to prevent mixing of the reagents prematurely. Record the temperature(s) of the solutions to the nearest 0.1 °C.

Add the HCl from the graduated cylinder all at once to the calorimeter, cover the cal-orimeter quickly, swirl the mixture for 30 seconds, and record the highest temperature reached by the mixture (to the nearest 0.1 °C).

Now reweigh and record the mass of your calorimeter and its contents. The mass of your solution, essentially water, is the difference between this mass and your original mass for the calorimeter.

From the change in temperature undergone by the mixture upon reaction, the total mass of the combined solutions and the specific heat capacity of water, 4.184 J/(g·°C), calculate the quantity of heat that flowed from the reactant species into the water of the solution.

Calculate the number of moles of $NH_4Cl(aq)$ produced when 50 mL of 1.5 M HCl reacts with 50 mL of 1.5 M NH_3.

Calculate ΔH_{neut} in terms of the number of kilojoules of heat energy transferred per 1 mol of $NH_4Cl(aq)$ formed in the neutralization of aqueous HCl with aqueous NH_3. Repeat the determination of ΔH_{neut} for the HCl/NH_3 reaction a second time as written above.

Part B: Heat of Solution

Calculate the weight of ammonium salt that was formed in the neutralization (Part A). Obtain this weight of the salt (nearest 0.01 g) on weighing paper. Carefully dry, weigh, and record the mass of your calorimeter (cups, lid, and thermometer). Add 100.0 mL of deionized water to the calorimeter, and record the premixing temperature. Raise the lid, dump the solid into the calorimeter, and quickly replace the lid. Swirl very frequently (to assist dissolution of the solid) while taking readings until a definite maximum or minimum temperature has been attained. (**Be patient**; dissolving a solid takes a little longer than the mixing of the two solutions.) Record a final temperature and then re-weigh your assembly to determine the mass of the solution you have created.

Calculate ΔH_{soln} for this reaction in kilojoules per mole of ammonium salt dissolved. Rinse and dry the calorimeter and the thermometer. Repeat a second time.

12.4 CALCULATIONS

Combine your average results for ΔH_{neut} and ΔH_{soln} with the $\Delta H_f°$ values for $NH_3(aq)$ and HCl(aq) to obtain a Hess's law sum that gives $\Delta H_f°$ for $NH_4Cl(s)$. Calculate the percent error between your answer and that tabulated by the NIST; that is:

$$\% \text{ error} = \frac{|\text{ NIST value} - \text{determined value }|}{\text{NIST value}} \times 100\%$$

Safety Notes

Everyone in the room will know if you do not keep the lid on the ammonia. Be careful of the HCl as it is a strong acid.

12.5 WASTE DISPOSAL

All solutions may be flushed down the drain with plenty of tap water.

EXPERIMENT 12: Hess's Law

DATA SHEET 1

Part A: Heat of Neutralization

	Trial 1	Trial 2
Experimental:		
Volume of 1.5 M NH_3	_____	_____
Volume of 1.5 M HCl	_____	_____
Mass of Dry Calorimeter	_____	_____
Initial Temperature of NH_3	_____	_____
Initial Temperature of HCl	_____	_____
Final Temperature Reached	_____	_____
Calculations:		
Average Initial Temperature	_____	_____
ΔT for Mixture	_____	_____
Total Mass of Calorimeter and Contents	_____	_____
Calculated Mass of Mixture	_____	_____
Heat Flow (joules)	_____	_____
Heat Flow (kilojoules)	_____	_____
Moles of NH_4Cl Produced	_____	_____
ΔH_{neut} of NH_4Cl*(aq)*	_____	_____

Avg. ΔH_{neut} _____

EXPERIMENT 12: Hess's Law

DATA SHEET 2

Part B: Heat of Solution

	Trial 1	**Trial 2**
Experimental:		
Mass of $NH_4Cl(s)$ Taken	_____	_____
Volume of H_2O Taken	_____	_____
Mass of Dry Calorimeter	_____	_____
Initial Temperature of H_2O	_____	_____
Final Temperature Reached	_____	_____
Total Mass of Calorimeter and Contents	_____	_____
Calculations:		
ΔT for Solution	_____	_____
Calculated Mass of Solution	_____	_____
Heat Flow (joules)	_____	_____
Heat Flow (kilojoules)	_____	_____
Moles of NH_4Cl Produced	_____	_____
ΔH_{soln} of $NH_4Cl(aq)$	_____	_____
Avg. ΔH_{soln}	_____	

Hess's Law Calculations:

Avg. ΔH_{neut} of $NH_4Cl(aq)$ from Part A	_____
Avg. ΔH_{soln} of $NH_4Cl(aq)$ from Part B	_____
Calculated ΔH_f° of $NH_4Cl(s)$	_____
% Error	_____

EXPERIMENT 12: Hess's Law

POSTLAB EXERCISE

1. Given

   $H_2(g) \rightarrow 2H(g)$ $\Delta H° = 436.4$ kJ
   $Br_2(g) \rightarrow 2Br(g)$ $\Delta H° = 192.5$ kJ
   $H_2(g) + Br_2(g) \rightarrow 2HBr(g)$ $\Delta H° = -104$ kJ

   Calculate the $\Delta H_f°$ for the reaction:

   $$H(g) + Br(g) \rightarrow HBr(g)$$

2. Given
   $C(graphite) + O_2(g) \rightarrow CO_2(g)$ $\Delta H° = -393.5$ kJ
   $C(diamond) + O_2(g) \rightarrow CO_2(g)$ $\Delta H° = -395.4$ kJ

   Calculate the standard enthalpy of formation for diamond.

3. Given the information results of question 2, calculate the heat change for the formation of 2.50 g of diamond.

EXPERIMENT 13: Let There Be Light

PRELAB EXERCISE

Terms:

    Electromagnetic radiation –

    Wavelength –

    Frequency –

    Photon –

    Planck's constant –

    Spectrophotometer –

Safety Warnings:

EXPERIMENT 13: Let There Be Light

PRELAB EXERCISE

1. What is the frequency of a photon of electromagnetic radiation with a wavelength of 2.64×10^{-7} m?

2. What is the energy of the photon of electromagnetic radiation from question 1?

13

LET THERE BE LIGHT

Objectives

1. To use a spectrophotometer to observe the colors of the visible light spectrum.

2. To calculate the energy of the colors of the visible light spectrum.

13.1 INTRODUCTION

Visible light is a form of electromagnetic radiation. Electromagnetic radiation is radiant energy that exhibits wavelike behavior and travels through space at the speed of light in a vacuum. The speed of light in a vacuum is a very fast 186,000 miles per second, or 3.00×10^8 m/s in the SI system.

Visible light has the same three characteristics as any other wave: wavelength, frequency, and speed. Wavelength (λ) is the distance between two identical consecutive points on a wave. Frequency (v) is the number of waves (called cycles) that pass a given point in space per second. The speed of a wave (c) is how fast that it travels. These three characteristics are related, as shown in Equation 1.

$$c = v\lambda \qquad \text{(Equation 1)}$$

Albert Einstein and Max Planck realized that visible light also acts like a particle. These "particles" of light are called photons. The energy of a photon is calculated using Equation 2, where h is Planck's constant, 6.626×10^{-34} J•s/photon.

$$E = hv \qquad \text{(Equation 2)}$$

Electromagnetic radiation covers quite a large range of wavelengths from the very long AM radio waves (~10^4 m) to the very short gamma rays (~10^{-12} m). The visible region of this range is quite small (~ 400 to ~700 nm). When combined, the colors of the visible spectrum form what we call white light.

In today's experiment we will take white light and use the monochromator (wavelength selector) in a spectrophotometer (device for measuring light) to isolate certain wavelengths for observation.

13.2 EXAMPLE CALCULATIONS

1. What is the frequency of a photon of electromagnetic radiation with a wavelength of 1.00 m?

$$\nu = \frac{c}{\lambda} \text{ giving } 3.00 \times 10^8 \text{ m/s} \div 1.00 \text{ m} = 3.00 \times 10^8 \text{s}^{-1}$$

2. What is the energy of the photon of electromagnetic radiation from question 1?

$$E = h\nu \text{ giving } (6.626 \times 10^{-34} \text{ J} \cdot \text{s/photon})(3.00 \times 10^8 \text{ s}^{-1}) = 1.99 \times 10^{-25} \text{ J/photon}$$

13.3 EXPERIMENTAL

Part A: The Colors of Visible Light

A piece of white chalk about 2 cm long should be rubbed to a 45° angle and placed in a cuvet as shown in Figure 2 on the first data sheet.

The spectrophotometer should be turned on and allowed to warm up. Select the Transmittance operating mode. Use the wavelength selector knob to set the wavelength to about 550 nm. Place the cuvet into the sample chamber with the angled side of the chalk facing right. It should look like Figure 1.

On Figure 2, draw an arrow (←) from the source of light to the cuvet at the level where you see the light beam hit the chalk. Rotate the cuvet until the band of light on the chalk is as wide as you can get it. Set the wavelength to 375 nm. Observe the color shown on the chalk and record it in Table 1. Repeat this process for all wavelengths at 25 nm intervals up to 725 nm. Answer the questions on Data Sheet 2.

Part B: Limits to Our Eyes

In Part A we found that the colors are hard to see below 400 nm and above 700 nm. In this part, we will look more closely at the visibility of the colors near the lower and upper limits of the visible light spectrum.

Set the wavelength to 375 nm. Observe the color (if any) shown on the chalk. Increase the wavelength by 5 nm intervals until you get the shortest wavelength at which you can see color. Record this wavelength and color on Data Sheet 3.

Set the wavelength to 675 nm. Observe the color (if any) shown on the chalk. Increase the wavelength by 5 nm intervals until you get the longest wavelength at which you can see color. Record this wavelength and color on Data Sheet 3.

Think of a way to adjust the brightness of the light outside the spectrophotometer. Apply your adjustment and repeat steps 1 and 2. Record your readings on Data Sheet 3 under Readings with Adjusted Light.

Figure 1 Spectrophotometer

13.4 CLEANUP

Turn off the spectrophotometer and make sure the sample compartment is empty. Place the cuvet and chalk in the labeled container. Clean up your work area. Wash your hands.

Safety Notes

Use general safety precautions.

13.5 WASTE DISPOSAL

None

EXPERIMENT 13: Let There Be Light

DATA SHEET 1

Part A: The Colors of Visible Light

Draw an arrow from the source of the light to the cuvet at the level where you see the light beam hit the chalk.

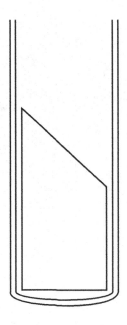

Figure 2 Cuvet with Chalk

EXPERIMENT 13: Let There Be Light

DATA SHEET 2

Part A: The Colors of Visible Light

Table 1 The Colors of Visible Light

Wavelength (nm)	Color
375	
400	
425	
450	
475	
500	
525	
550	
575	
600	
625	
650	
675	
700	
725	

1. Which color has the widest range?

2. What wavelength range does this color cover?

3. What percentage of the spectrum does this color occupy (use 375 to 725 nm for the range of visible light)?

4. In what part of the spectrum do the colors change fastest?

5. What colors are seen there?

EXPERIMENT 13: Let There Be Light

DATA SHEET 3

Part B: Limits to Our Eyes

1. Initial Readings

    a. Shortest wavelength at which color is visible: _____ nm

    b. Color seen: _____

    c. Longest wavelength at which color is visible: _____ nm

    d. Color seen: _____

2. Readings with Adjusted Light

    a. Shortest wavelength at which color is visible: _____ nm

    b. Color seen: _____

    c. Longest wavelength at which color is visible: _____ nm

    d. Color seen: _____

Method used to adjust the light around the spectrophotometer:

EXPERIMENT 13: Let There Be Light

POSTLAB EXERCISE

1. Your lab instructor will assign a color. Write that color here: _____

2. Based upon your results, what wavelength in nanometers is the color you were assigned?

3. Based upon your results, what frequency in hertz (Hz) is the color you were assigned?

4. What energy in joules is a photon of light of the color you were assigned?

EXPERIMENT 14: Paper Chromatography

PRELAB EXERCISE

Terms:

Chromatography –

Chromatogram –

Moving phase –

Stationary phase –

Gas chromatography –

Thin-layer chromatography –

Paper chromatography –

Safety Warnings:

EXPERIMENT 14: Paper Chromatography

PRELAB EXERCISE

1. From a mixture, on what basis can the substances be separated using the technique of chromatography?

14 PAPER CHROMATOGRAPHY

Objectives

1. To gain experience in chromatography technique.

2. To determine the components in a mixture through chromatography.

3. To identify an unknown through comparison of knowns on a chromatogram.

14.1 INTRODUCTION

Chromatography is a technique used to separate and identify small quantities of mixtures into their component parts. This technique was so named due to the highly colored (from the Greek *chroma*, meaning "colored") components separated in 1906 when chloroplast extracts were separated with this technique.

All forms of chromatography employ two phases, a *moving phase* and a *stationary phase*. The moving phase is a solvent, or mixture of solvents, that flows over the other material. In chromatography, mixtures can be separated because each substance in the mixture will have different affinities for the stationary and moving phases. When the correct stationary and moving phases are selected, the mixture separates because:

> *each substance in the mixture is adsorbed on the stationary phase with a different degree of tenacity; and*
> *each substance in the mixture has a different affinity for the moving phase.*

A compound with a high affinity for the moving phase and a low affinity for the stationary phase will move through the *stationary phase* rapidly. A compound with a high affinity for the stationary phase and a low affinity for the moving phase will move through the *stationary phase* slowly.

There are several types of chromatography in use. Gas chromatography utilizes a gas as the moving phase. With thin-layer chromatography (TLC) the adsorbent layer, stationary phase, is spread in a thin-layer on a plastic or glass plate. In paper chromatography the stationary phase is a paper similar to filter paper. Although paper chromatography is not as effective as gas or thin-layer chromatography, it is fast, inexpensive, and can give good results.

In TLC and paper chromatography the mobility of a substance in a mixture is stated in terms of its R_f value. R_f value is a physical property of a substance. R_f value equals the ratio of the distance traveled by the substance divided by the distance traveled by the solvent. Each distance is measured from the point at which the mixture is applied to the adsorbent.

$$R_f \text{ Value} = \frac{\text{distance traveled by the substance}}{\text{distance traveled by the solvent}}$$

The R_f value is calculated as follows:

Measure the distance traveled by the solvent in millimeters from the initial position of the spot to the position marked at the edge of the solvent front.

Measure the distance in millimeters traveled by each colored component from the initial position of the spot to the center of each spot.

Divide the "component distance" by the "solvent distance" for each spot.

Whether the component spot travels 15 mm or 40 mm from the starting point, the ratio of that distance to the distance traveled by the solvent will remain constant. Under the same set of conditions (i.e., same solvent and the same type of paper), a compound will have the same R_f value whenever it is analyzed. Because of this fact R_f values can be used to identify compounds.

14.2 EXPERIMENTAL

Obtain:

three 250 mL beakers

three 6 x 8 cm chromatography papers

plastic wrap

metric ruler

toothpicks

1. Clean the 250 mL beakers. Label them "deionized water," "rubbing alcohol," and "0.10% NaCl." These will be your chromatography chambers.

2. The rubbing alcohol reagent container is located on the side-shelf. A labeled 50 mL beaker and a labeled 10 mL graduated cylinder are located beside the rubbing alcohol reagent container. Bring your 250 mL "rubbing alcohol" chromatography chamber to the rubbing alcohol reagent container. Pour from the rubbing alcohol reagent container into the labeled 50 mL beaker already provided. From the provided 50 mL beaker, pour 7 mL of rubbing alcohol into the graduated cylinder. Pour the liquid from the graduated cylinder into the 250 mL rubbing alcohol chromatography chamber. Cover the rubbing alcohol chromatography chamber with plastic wrap.

3. The deionized water reagent container is located on the side-shelf. A labeled 50 mL beaker and a labeled 10 mL graduated cylinder are located beside the deionized water reagent container. Bring your 250 mL "deionized water" chromatography chamber to the deionized water reagent container. Pour from the deionized water reagent container into the labeled 50 mL beaker already provided. From the provided 50 mL beaker, pour 7 mL of deionized water into the graduated cylinder. Pour the liq-

uid from the graduated cylinder into the 250 mL deionized water chromatography chamber. Cover the deionized water chromatography chamber with plastic wrap.

4. The 0.10% NaCl reagent container is located on the side shelf. A labeled 50 mL beaker and a labeled 10 mL graduated cylinder are located beside the 0.10% NaCl reagent container. Bring your 250 mL "0.10% NaCl" chromatography chamber to the 0.10% NaCl reagent container. Pour from the 0.10% NaCl reagent container into the labeled 50 mL beaker already provided. From the provided 50 mL beaker, pour 7 mL of 0.10% NaCl into the graduated cylinder. Pour the liquid from the graduated cylinder into the 250 mL 0.10% NaCl chromatography chamber. Cover the 0.10% NaCl chromatography chamber with plastic wrap.

5. While the vapors from the solvent are equilibrating in the development chamber, take three of the chromatography papers and place them on a clean paper towel making sure to handle them by one edge at all times. On each paper draw a line, in pencil, 1.5 cm from the bottom across the short edge. Repeat this on the top of the paper. On the bottom line, evenly space six small x's along the line (do not place an x near the edge). DO NOT USE INK ON THE CHROMATOGRAM. (Ink is soluble in the solvents and will itself migrate on the chromatogram.) Any labeling should be done in pencil. Place a letter code under each small x for each of the colors; R for red, as an example. Place an ID for each of the chromatography chambers so you will be able to tell them apart after the experiment.

6. The lab instructor will prepare individual labeled well plates with each dye and unknown present in only one well plate. Use the small end of the toothpick provided to spot each dye and unknown onto each of the three pieces of chromatography paper. To spot each dye, place one end of the toothpick in the dye, and allow the toothpick to soak up some of the dye. Then, keeping the toothpick vertical, touch the toothpick to the mark of the origin line on your paper. Keep your spots in as small an area as possible. Before respotting, allow the spot from each individual application to dry. Spot the dyes on their appropriate marks. The unknowns may be much less concentrated than your knowns. You should respot them until the intensity of their color is near that of the knowns. Remember to allow the spot to dry before respotting.

7. Uncover the 250 mL chromatography chamber and quickly place the appropriate chromatography paper inside it with the bottom edge (where the dyes were applied) touching the solvent. The edges of the paper will touch the sides of the beaker. Re-cover the beaker with the plastic wrap. Let the solvent move to the top line on the paper. After the solvent has reached this point, remove the chromatography paper and quickly mark where the solvent front stopped. Draw an ellipse around the edge of each color. Gently wave the chromatography paper to assist its drying.

8. After the chromatography paper has dried, measure the distance the solvent moved in millimeters.

9. Find the center of the ellipse of each color. Use the original marking upon removing the paper from the chromatography chamber even if the spots moved as the chromatography paper dried. Measure the distance the spot has traveled in millimeters.

10. Calculate the R_f for each dye in each mobile phase. Your unknowns may have more than one dye.

11. Save your chromatograms to turn in with your lab report.

Safety Notes

Use general safety precautions.

14.3 WASTE DISPOSAL

Pour the rubbing alcohol into the "reclaimed rubbing alcohol" container. Other liquid waste for this lab may be dissolved in tap water and flushed down the sink.

NAME: _____ LAB SECTION: _____

UNKNOWN 1: _____

UNKNOWN 2: _____

EXPERIMENT 14: Paper Chromatography

DATA SHEET 1

Chromatogram Analysis

Complete the following three tables.

Solvent:_____ Original Color	Color	Distance Dye Traveled (mm)	Distance Solvent Traveled (mm)	R_f
Unknown 1				
Unknown 2				

EXPERIMENT 14: Paper Chromatography

DATA SHEET 2

Solvent:_____ Original Color	Color	Distance Dye Traveled (mm)	Distance Solvent Traveled (mm)	R_f
Unknown 1				
Unknown 2				

Solvent:_____ Original Color	Color	Distance Dye Traveled (mm)	Distance Solvent Traveled (mm)	R_f
Unknown 1				
Unknown 2				

EXPERIMENT 14: Paper Chromatography

POSTLAB EXERCISE

1. Based upon R_f values, which solvent would you expect to give the best separation of the unknowns?

2. Did the solvent selected in question 1 give the best separation? Explain why.

3. Which dyes are present in Unknown 1? How do you know?

4. Which dyes are present in Unknown 2? How do you know?

EXPERIMENT 15: Lewis Structures, VSEPR, and Molecular Shape

PRELAB EXERCISE

Terms:

    Lewis structure –

    Valence electrons –

    VSEPR –

    VSEPR model –

    Lone pair electrons –

    Bond pair electrons –

    Formal charge –

    Resonance –

Safety Warnings:

EXPERIMENT 15: Lewis Structures, VSEPR, and Molecular Shape

PRELAB EXERCISE

1. Draw the Lewis structure for CH_3Cl and assign formal charges to each atom.

2. Draw the Lewis structure for PCl_5 and assign formal charges to each atom.

3. In the Molymod molecular model kit, carbon is represented by which two colors?

4. In the Molymod molecular model kit, a yellow ball represents which element?

5. Draw the preferred Lewis structures for the molecules on the first three data pages of this lab.

15

LEWIS STRUCTURES, VSEPR, AND MOLECULAR SHAPE

Objectives

1. To gain experience drawing Lewis structures.

2. To learn to use VSEPR theory to predict molecular shapes.

3. To learn how to construct molecular models using a model kit.

15.1 INTRODUCTION

Lewis Structure

The *Lewis structure* is a representation that shows the atoms present in a molecule and how the *valence electrons* (those in the outermost principal quantum level of an atom and any unfilled inner levels) are arranged among these atoms. Two important concepts are combined in determining Lewis structures. First, valence electrons play a fundamental role in chemical bonding. This makes sense since they will be the first things in an atom that another species will "sense" as it approaches the atom to bond with it. Second, a noble gas electronic configuration is low in energy. In other words, atoms will combine to form molecules in such a way as to obtain a noble gas configuration.

In drawing Lewis structures, the nucleus and inner shell electrons are represented by the element's symbol. The valence electrons may be either represented individually by dots or in groups of two by a dash. The fluorine molecule is shown as an example in Figure 1.

Figure 1 Lewis structure of the fluorine molecule

Note that the 1s electrons of each fluorine atom are not included in the structure.

There are some rules for drawing Lewis structures.

1. Determine the formula of the molecule.

2. Determine the total number of valence electrons.

3. Identify the central atom(s). This is usually the one that can form the most bonds to obtain a noble gas configuration. (It is never hydrogen.)

4. Draw an outline structure of the atoms and connect the atoms by single bonds.

5. Use the remaining electrons to complete noble gas configurations for exterior atoms and then the central one(s).

6. If the central atom still lacks a noble gas configuration, form multiple bonds.

Exceptions to the Noble Gas Rule

There are three exceptions to the noble gas rule. First, odd-electron species. All noble gases have an even number of electrons. There is no way to pair an odd number of electrons. Second, incomplete noble gas configurations. These are Be, B, and Al species. Third, expanded noble gas configurations. These are usually nonmetal atoms of the third period or beyond bound to highly electronegative atoms.

Formal Charge and Selecting a Preferred Lewis Structure

Formal charge is the number of valence electrons in the free neutral atom minus the number of electrons assigned to that atom in the Lewis structure. A simple equation for determining formal charge is given:

$$FC = VE_{freeatom} - \# \text{ nonbonding e}^- - \# \text{ bonds}$$

where FC is the formal charge of the atom in question, $VE_{freeatom}$ is the number of valence electrons in the free atom, # nonbonding e^- is the number of nonbonding electrons assigned to the atom in the Lewis structure, and # bonds is the number of bonds that the atom has in the structure. The sum of the formal charges for a structure will equal the overall charge on the structure.

When more than one Lewis structure is possible for a compound, formal charges are used to determine the preferred structure. The process is as follows:

1. Draw the possible Lewis structures.

2. Assign formal charges to all atoms in the possible Lewis structures.

3. Preferred structures will have negative formal charges assigned to more electronegative atoms in the structure. Positive formal charges will be assigned to less electronegative atoms in the structure.

4. If step 3 does not yield a preferred structure, select the structure that minimizes formal charges.

Two possible Lewis structures for nitric acid are shown in Figure 2. Let us use the rules above to determine the preferred structure.

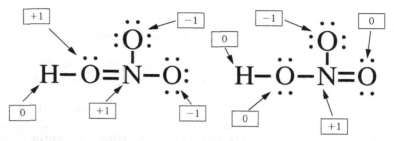

Figure 2 Two possible Lewis structures for nitric acid

The formal charges have been added to the structures in Figure 2. The structure on the right is preferred for two reasons. First, oxygen does not have to take a +1 oxidation state. It is the most electronegative atom in nitric acid (Rule 3). Second, there are fewer atoms assigned an oxidation state in the structure on the right (Rule 4).

Resonance

Sometimes there is more than one Lewis structure possible and formal charge gives no preference. This is the situation of *resonance*. An example of resonance is found in nitric acid. Look at the right side of Figure 2. Is there any reason the oxygen to the right of the nitrogen should get the double bond over the oxygen above the nitrogen? Of course there is not. Nitric acid has a resonance, as shown in Figure 3, with the double bond alternating between these two oxygens.

$$\text{H}-\overset{..}{\underset{..}{\text{O}}}-\overset{\overset{..}{\text{O}}..}{\underset{\vphantom{.}}{\text{N}}}-\overset{..}{\underset{..}{\text{O}}}: \;\longleftrightarrow\; \text{H}-\overset{..}{\underset{..}{\text{O}}}-\overset{\overset{..}{\overset{..}{\text{O}}}..}{\underset{\vphantom{.}}{\text{N}}}=\overset{..}{\underset{..}{\text{O}}}$$

Figure 3 Resonance of nitric acid

VSEPR

Once the Lewis structure of a molecule is determined, the arrangement of valence electrons can be used to determine the molecular shape using valence-shell electron-pair repulsion (VSEPR) theory. The ideas used in VSEPR theory are as follows:

1. Valence pairs of electrons repel each other since they have like charges.

2. Electron pairs orient themselves about an atom so as to minimize these repulsions.

3. The closer together two groups of electrons are, the stronger the repulsion between them.

4. *Lone pair electrons* (valence electrons around an atom not involved in bonding with other atoms) spread out more than *bond pair electrons* (valence electrons involved in bonding between atoms).

Determining the molecular shape around an atom of interest is outlined in the following rules:

1. Determine the number of electron groups around the atom of interest. Lone pairs of electrons count as one electron group each. Multiple bonds to the same atom only count once.

2. The total number of electron groups determines the electron group geometry (column 2 of the molecular geometries table).

3. The electron group geometry is modified by the number of lone pairs around the atom of interest (column 3) to generate the molecular shape (column 4).

Molecular Geometries

Number of Electron Groups	Electron Group Geometry	Number of Lone Pairs	Molecular Shape	Ball and Stick Structure	Generic Formula
2	linear	0	linear		AB_2
3	trigonal planar	0	trigonal planar		AB_3
3	trigonal planar	1	bent		AB_2
3	trigonal planar	2	linear		AB
4	tetrahedral	0	tetrahedral		AB_4
4	tetrahedral	1	trigonal pyramidal		AB_3
4	tetrahedral	2	bent		AB_2
4	tetrahedral	3	linear		AB
5	trigonal bipyramid	0	trigonal bipyramid		AB_5
5	trigonal bipyramid	1	see-saw		AB_4
5	trigonal bipyramid	2	T-shape		AB_3
5	trigonal bipyramid	3	linear		AB_2
5	trigonal bipyramid	4	linear		AB
6	octahedral	0	octahedral		AB_6
6	octahedral	1	square pyramid		AB_5
6	octahedral	2	square planar		AB_4
6	octahedral	3	T-shape		AB_3
6	octahedral	4	linear		AB_2
6	octahedral	5	linear		AB

VSEPR Model

In order to represent on paper the three-dimensional structures that we will be looking at in lab, we will combine Lewis structures and VSEPR theory to create VSEPR models. The modifications are as follows:

1. An atom is selected to be the atom of interest. The structure around this atom is what is drawn.

2. Lone pair electrons on other atoms are ignored.

3. The central atom is in the plane of the paper. The structure is drawn in such a way that the maximum number of atoms is in the plane of the paper.

4. Atoms coming out of the plane of the paper to you are represented by a line pointing at the central atom and getting thicker as it moves away from the central atom to the atom coming out of the paper.

5. Atoms going into the plane of the paper are represented by a dashed (dotted) line going from the central atom to the atom going into the plane of the paper.

An example of the tetrahedral structure of methane, CH_4, is shown in Figure 4 as ball and stick and VSEPR model.

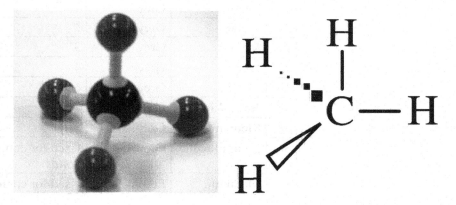

Figure 4 Ball and stick and VSEPR model of methane

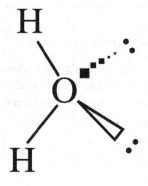

Figure 5 The VSEPR model of water

The central carbon and the two hydrogens on the right are in the plane of the paper. The hydrogen in the upper left is going into the plane of the paper and the one in the lower left is coming straight out of the paper.

What about lone pairs? The electron group structure remains but now the lone pair electrons replace some terminal atoms. An example of water is shown in Figure 5.

For the VSEPR model of water, the electron group structure is tetrahedral. The maximum number of atoms is placed in the plane of the paper, all three in this case. One lone pair of electrons is going into the plane of the paper and the other is coming out of the plane of the paper.

Molymod Molecular Model Kit: Each molecular model kit contains the items listed in the following table.

No.	Typical Element	Ball Color	Holes	Electron Group Geometry
14	Carbon	Black	4	Tetrahedral
6	Carbon	Dk blue	5	Trigonal bipyramidal
12	Hydrogen	White	1	
2	Hydrogen	White	2	Linear
6	Nitrogen	Blue	4	Tetrahedral
4	Nitrogen	Blue	3	Trigonal pyramid
13	Oxygen	Red	2	Bent
4	Oxygen	Red	4	Tetrahedral
5	Oxygen	Red	1	Linear
8	Sulfur	Yellow	2	Bent
4	Sulfur	Yellow	4	Tetrahedral
1	Sulfur	Yellow	6	Octahedral
4	Phosphorus	Purple	4	Tetrahedral
1	Phosphorus	Purple	5	Trigonal bipyramid
2	Phosphorus	Purple	3	Trigonal pyramid
8	Halogen	Green	1	
4	Metal	Grey	1	
3	Metal	Grey	2	Bent
2	Metal	Grey	3	Trigonal pyramid
4	Metal	Grey	4	Tetrahedral
1	Metal	Grey	6	Octahedral
No.	Bond Links	Color		
38	Medium	Grey		
36	Long Flexible	Grey	Used for double bonds	
12	Medium	Purple	Used for contrast	

15.2 EXPERIMENTAL

Verify the preferred Lewis structure for each of the species listed in the data tables.

Your lab instructor will assign which structures to determine the molecular geometry. Your lab instructor will inform you which atom to use in determining the molecular geometry in some instances. Construct the molecular model for the structure and draw the VSEPR model for each of the species listed on the data tables. For part D, your lab instructor will assign four species as your unique unknowns.

15.3 CLEANUP

There will be hundreds of students using the model kits during the year. Make sure to return the parts for the model kit to where you found them.

EXPERIMENT 15: Lewis Structures, VSEPR, and Molecular Shape

DATA SHEET 1

Part A: Compounds with Single Bonds

Compound	Lewis Structure	Molecular Geometry	VSEPR Model
NH_3			
CCl_4			
PCl_5			
SF_6			
H_2Se			

EXPERIMENT 15: Lewis Structures, VSEPR, and Molecular Shape

DATA SHEET 2

Part B: Compounds with Double and Triple Bonds

Underlined atoms are the atoms of interest.

Compound	Lewis Structure	Molecular Geometry	VSEPR Model
CO_2			
$\underline{C}_2H_4$			
N_2			
HO$\underline{C}$N			

EXPERIMENT 15: Lewis Structures, VSEPR, and Molecular Shape

DATA SHEET 3

Part C: Acids and Polyatomic Ions

Underlined atoms are the atoms of interest.

Species	Lewis Structure	Molecular Geometry	VSEPR Model
OH^-			
NH_4^+			
ClO_3^-			
$H_2\underline{S}O_4$			
Acetic Acid $CH_3\underline{C}O_2H$			

EXPERIMENT 15: Lewis Structures, VSEPR, and Molecular Shape

DATA SHEET 4

Part D: Unknowns (3 points each)

Species	Lewis Structure	Molecular Geometry	VSEPR Model

EXPERIMENT 15: Lewis Structures, VSEPR, and Molecular Shape

POSTLAB EXERCISE

1. Draw Lewis structures of the two resonance structures of ozone (O_3) and determine the formal charges for each atom to show that they are equivalent.

2. Draw the Lewis structure of XeF_2 and assign formal charges to all atoms.

3. What is the molecular shape of the species in question 2?

4. Draw the VSEPR model of the species in question 2.

EXPERIMENT 16: Intermolecular Forces

PRELAB EXERCISE

Terms:

Intermolecular forces –

London dispersion force –

Polar molecule –

Dipole–dipole attraction –

Hydrogen bonding –

Safety Warnings:

EXPERIMENT 16: Intermolecular Forces

PRELAB EXERCISE

Use the periodic table provided in your lab manual to calculate the molar mass of each compound in the table on Data Sheet 1. Using the structures provided in the table, determine whether or not each compound is polar and if hydrogen bonding is present. Record your results in the corresponding columns of the table prior to coming to lab. The remainder of the table will be completed using experimental data.

16

INTERMOLECULAR FORCES

Objective

To study the relationship between intermolecular forces and ΔT_{max}.

16.1 INTRODUCTION

When a liquid evaporates from a surface, it draws energy from the surface, causing the surface to cool. Like boiling point and viscosity, the magnitude of the temperature change that occurs is related to the strength of the intermolecular forces of attraction in the liquid. As the strength of attractions between molecules increases, the evaporation rate decreases, resulting in less of a temperature change. You will use your results to compare the intermolecular forces in sets of related compounds.

The classes of organic solvents studied in this experiment are alkanes, alcohols and, for more thorough investigation, a ketone. Alkanes are hydrocarbons composed of single bonds between carbon atoms. Alcohols are hydrocarbon derivatives in which a hydroxyl functional group (-OH) is attached to a carbon atom in the compound. Ketones are hydrocarbon derivatives with two carbon atoms bonded to the carbon in a carbonyl functional group—that is, an oxygen atom double-bonded to a carbon atom. You will examine the molecular structure of these compounds, looking for the presence of hydrogen bonding, dipole–dipole interactions, and dispersion forces, all of which are types of intermolecular forces. Recall that polarity (including compounds that hydrogen bond) contributes to the strength of intermolecular forces as do increasing molecular mass (because of dispersion forces) and structural elongation.

16.2 EXPERIMENTAL (WORK IN PAIRS)

Your lab instructor will provide you and your partner with the following:

*   digital thermometer
*   stopwatch
*   filter paper
*   ruler
*   pair of scissors
*   1–2 inch piece of wire
*   set of 10 capped 13 mm test tubes, each containing a different organic solvent

Using a pencil (solvents can dissolve ink) and the ruler provided, draw 10 rectangles measuring 2.5 cm x 5 cm on filter paper in such a way as to fit as many rectangles on a single piece of paper as possible. You may alter the dimensions *slightly* if by doing so more rectangles can be drawn on the paper, but for best results, it is important that the paper strips are the same size. A template may be provided for you. Carefully cut out the rectangles, reserving any paper large enough to be used to make more of these rectangles. Discard the smaller scraps.

Tightly wrap a strip of paper in a cylinder around the tip of the thermometer probe, starting with the short edge of the strip parallel to the probe. Be sure to keep the long edges aligned as you wrap. Secure the strip *tightly* around the probe with the wire. Use your fingers to flatten the tips of the wire against the paper-covered probe as much as possible. The end of the paper cylinder should be flush with the end of the thermometer probe tip.

Under a fume hood, uncap the test tube containing the methanol. Dip the wrapped probe tip of the thermometer in the methanol. When it has been in the liquid for at least 30 seconds, turn on the thermometer to obtain the initial temperature of the liquid. Record the temperature to the nearest 0.1 °C on Data Sheet 2 at Time = 0.00. Working quickly, remove the thermometer from the test tube, place it on the lab bench under the fume hood as demonstrated by your lab instructor, then *immediately recap the test tube*. (Should the paper and wire fall into the solvent, leave them in the solvent and report it to your lab instructor.) Record the temperature to the nearest 0.1 °C every 15 seconds until a minimum is reached for 3 consecutive readings, or until 3 consecutive higher temperatures are reached. For some compounds, it may take longer than 4 minutes to reach the minimum temperature, in which case you should keep recording the temperatures. **Note: There may be small fluctuations in the temperature as the liquid evaporates from the paper, but if the temperature begins to rise consistently, it is unnecessary to continue recording the temperature.**

Repeat the procedure with the remaining solvents, using a fresh paper strip for each solvent but reusing the same piece of wire. You should see trends in ΔT_{max} within related groups of solvents. *If you obtain a result for a solvent that does not fit the expected trend, consult your lab instructor for further instructions.* When you have collected all necessary data, calculate the maximum change in temperature, ΔT_{max}, for each compound by subtracting the lowest recorded temperature from the initial temperature. Be sure to transfer your values into the table in Data Sheet 1.

Safety Notes

The compounds being used in today's experiment are toxic and flammable. Avoid inhalation and contact with the eyes and skin. There should be no open flame in the lab.

16.3 WASTE DISPOSAL

Make sure the test tubes containing the solvents are tightly capped. Return the tubes in the test tube rack to the large fume hood in the lab. Place only filter paper that has been exposed to a solvent and all broken copper wire in designated waste containers. Paper scraps that are too small to cut out at least one full rectangle should be placed in the regular trash. Larger pieces of filter paper, and all wire that is not broken, should be returned to the main bench for reuse.

EXPERIMENT 16: Intermolecular Forces

DATA SHEET 1

Solvent Information

Liquid	Structure	Molecular Weight	Polar? (yes/no)	H-bonding? (yes/no)	ΔT_{max}
Methanol	CH_3-OH				
Ethanol	CH_3CH_2-OH				
1-Propanol	$CH_3CH_2CH_2-OH$				
2-Propanol	CH_3CHCH_3 | OH				
1-Butanol	$CH_3CH_2CH_2CH_2-OH$				
Acetone	CH_3CCH_3 $\parallel$ O				
n-Pentane	$CH_3(CH_2)_3CH_3$				
n-Hexane	$CH_3(CH_2)_4CH_3$				
n-Heptane	$CH_3(CH_2)_5CH_3$				
n-Octane	$CH_3(CH_2)_6CH_3$				

EXPERIMENT 16: Intermolecular Forces

DATA SHEET 2

Experimental Temperature Data (°C) for Various Alcohols

Time (min:sec)	Methanol Temp.	Ethanol Temp.	1-Propanol Temp.	2-Propanol Temp.	1-Butanol Temp.
0:00	_____	_____	_____	_____	_____
0:15	_____	_____	_____	_____	_____
0:30	_____	_____	_____	_____	_____
0:45	_____	_____	_____	_____	_____
1:00	_____	_____	_____	_____	_____
1:15	_____	_____	_____	_____	_____
1:30	_____	_____	_____	_____	_____
1:45	_____	_____	_____	_____	_____
2:00	_____	_____	_____	_____	_____
2:15	_____	_____	_____	_____	_____
2:30	_____	_____	_____	_____	_____
2:45	_____	_____	_____	_____	_____
3:00	_____	_____	_____	_____	_____
3:15	_____	_____	_____	_____	_____
3:30	_____	_____	_____	_____	_____
3:45	_____	_____	_____	_____	_____
4:00	_____	_____	_____	_____	_____
ΔT_{max} ($T_{initial}$ - T_{lowest})	_____	_____	_____	_____	_____

EXPERIMENT 16: Intermolecular Forces

DATA SHEET 3

Experimental Temperature Data (°C) for a Ketone and Various Alkanes

Time (min:sec)	Acetone Temp.	n-Pentane Temp.	n-Hexane Temp.	n-Heptane Temp.	n-Octane Temp.
0:00	___	___	___	___	___
0:15	___	___	___	___	___
0:30	___	___	___	___	___
0:45	___	___	___	___	___
1:00	___	___	___	___	___
1:15	___	___	___	___	___
1:30	___	___	___	___	___
1:45	___	___	___	___	___
2:00	___	___	___	___	___
2:15	___	___	___	___	___
2:30	___	___	___	___	___
2:45	___	___	___	___	___
3:00	___	___	___	___	___
3:15	___	___	___	___	___
3:30	___	___	___	___	___
3:45	___	___	___	___	___
4:00	___	___	___	___	___

ΔT_{max} ($T_{initial} - T_{lowest}$) ___ ___ ___ ___ ___

EXPERIMENT 16: Intermolecular Forces

POSTLAB EXERCISE

Interpretation of Data

For each solvent set below, use the information you have obtained to concisely explain variations in the maximum change in temperature based on differences in the relative strengths of intermolecular forces within a given set.

1. Alcohols:

2. Alkanes:

3. 2-propanol and acetone:

4. 1-butanol, n-pentane:

EXPERIMENT 17: Concentration Units

PRELAB EXERCISE

Terms:

      Molarity –

      Molality –

      % (m/m) –

      Volumetric unit –

      Gravimetric unit –

Safety Warnings:

EXPERIMENT 17: Concentration Units

PRELAB EXERCISE

1. In the titration labs you used the weak acid potassium hydrogen phthalate (formula $KHC_8H_4O_4$) as a primary standard in an acid–base titration. You may remember that the acid is commonly referred to as "KHP." How much of this weak acid would have to be weighed out to prepare 100.0 mL of a 0.240 M KHP solution?

2. If you actually weighed out 8.50% more KHP than required, what would be the actual molarity of the solution?

3. If you used the prepared KHP solution in a lab experiment, which concentration would you need to use in your calculations? Explain your reasoning.

4. You weigh out 2.98 g of sodium sulfate and use it to create a 250 mL aqueous solution. What is the molarity of sodium ions in the resulting solution?

17

CONCENTRATION UNITS

Objectives

1. To gain experience preparing solutions of known concentrations.

2. To learn about concentration units.

3. To gain experience in preparing labels.

17.1 INTRODUCTION

You will gain experience in preparing solutions and the calculations involved in preparing solutions of known concentrations. You will be using three different ways of expressing concentration.

$$\text{Molarity (M)} = \text{moles}_{solute} / \text{liters}_{solution}$$

$$\text{Molality } (m) = \text{moles}_{solute} / \text{kg}_{solvent}$$

$$\% \text{ (m/m)} \quad = 100\% \times \text{grams}_{x} / \text{total mass}$$

Molality and % (m/m) are examples of gravimetric units. Their value is set by mass. Molarity is an example of a volumetric unit. Its value is determined by volume.

In any area of science, three very important lab techniques to learn are careful weighing, careful solution preparation, and labeling of solutions. This experiment is especially designed to give you hands-on experience with the all of these. Your assignments are summarized as follows:

Labels

You will prepare labels for the solutions for each part of this week's lab on Data Sheets 3 and 4. The label will include the concentration (use the requested units for that part), the solute, the date prepared, and the preparer's initials. Your instructor can help you with an example. Remember that the label should be for the solution as prepared.

Part A

Assignment — Preparation of 100.00 mL of an aqueous solution of sucrose, $C_{12}H_{22}O_{11}$, with a specified molarity

Calculation — Mass of sucrose required

Part B

<u>Assignment</u> — Dilution of a 1.0 M NaCl solution to a specified final volume

<u>Calculation</u> — Final molarity of NaCl

Part C

<u>Assignment</u> — Preparation of 100.00 mL of solution with a specified mass of alum, $KAl(SO_4)_2 \cdot 12H_2O$. Student will determine the molarity of a particular ion (either K^+, Al^{3+}, or SO_4^{2-}) in the resulting solution (either K^+, Al^{3+}, or SO_4^{2-}).

<u>Calculation</u> — Molarity of ion in the final solution

Part D

<u>Assignment</u> — Preparation of a sucrose solution at a specified % (m/m) sucrose and a specified total mass of solution

<u>Calculation</u> — Masses of sucrose and water required

Part E

<u>Assignment</u> — Preparation of a solution with specified molality of sucrose using a specified mass of water

<u>Calculation</u> — Masses of sucrose required and solution

NOTE: For **Parts A, B, and C,** tap water will be used as the solvent, for reasons of expense. (Distilled or deionized water should always be used when the solutions are to be used in subsequent experiments.) Please note that molarity is a volumetric unit (based on volume of solution), while mass percentage and molality are gravimetric units (based on masses of solute and solvent, and hence, mass of solution).

Additional Note: Parts D and E are dry labs. That means you will only do the calculations.

Part F

<u>Assignment</u> — Prepare a standard solution of KIO_3 for Experiment 18

<u>Calculation</u> — Molarity of prepared solution

NOTE: For **Part F,** it is important to *use deionized water*, since this solution will be used in the next experiment as a standard.

Instructor Signoff

You must show your solutions for Parts A, C, and F to your instructor and have the instructor sign off on your prep work.

17.2 Experimental

Part A: Preparation of a Molar Solution

Your instructor will assign you a molarity of sucrose ($C_{12}H_{22}O_{11}$) solution, which you are to prepare in a 100.00 mL volumetric flask. Record the assigned molarity, the volume of solution (100 mL), and the calculated mass of sucrose needed. Transfer the sucrose to the volumetric flask and use your wash bottle to rinse any solid crystals adhering to the sides of the flask down into the bottom of the flask. Half-fill the flask with water and completely dissolve the sucrose. Fill the flask to exactly the 100 mL calibration mark. Cover the top of the flask with parafilm. Invert the flask 30 times, as demonstrated by your instructor, to thoroughly mix the contents. Include on your data sheet a diagram of a label that would be placed on this solution to identify it. The label should include the concentration of the solution, the solute, the date prepared, and initials of the person who prepared the solution. Discard the solution after consulting with your instructor on your method of preparation. (Since solution volume is temperature-dependent, molarity will vary somewhat with temperature.)

Volumetric flask

Part B: Dilution of a "Stock Solution"

A stock solution is one that has been carefully prepared with known molarity. Obtain 10.0 mL of 1.0 M NaCl stock solution in a 100 mL graduated cylinder. Record this volume. Dilute this sample with water to the final volume assigned by your instructor. What is the new molarity of the diluted solution? Record the new molarity on your data sheet and on the diagrammed label.

Part C: Preparation of a Solution of a Strong Electrolyte

You will use $KAl(SO_4)_2 \cdot 12H_2O$ [Molar Mass = 474.39 g/mol]. Your instructor will assign you a mass range of alum. Weigh out a mass of alum within this range. Transfer the alum to the volumetric flask and use your wash bottle to rinse any solid crystals adhering to the sides of the flask down into the bottom of the flask. Half-fill the flask with water and completely dissolve the alum. Fill the flask to exactly the 100 mL calibration mark. Cover the top of the flask with parafilm. Invert the flask 30 times, as demonstrated by your instructor, to thoroughly mix the contents. Include on your data sheet a diagram of a label that would be placed on this solution to identify it. The label should include the concentration of the solution, the solute, the date prepared, and initials of the person who prepared the solution. Discard the solution after consulting with your instructor on your method of preparation. (Since solution volume is temperature-dependent, molarity will vary somewhat with temperature.) Your instructor will give you a specific ion. Calculate the molarity of that ion in the solution that you just prepared.

Part D: Preparation of a Solution with Specific Percent by Mass

The instructor will assign a certain mass of solution and specific percent by mass of sucrose. Calculate and record the masses of sucrose and water, and include a drawing of a solution label, as before. Note: Mass percent is temperature-independent.

Part E: Molality

The instructor will assign you a mass of water to use and a specified molality of sucrose. Calculate and record the mass of sucrose needed. Provide a drawing of a label for this solution, as before. Note that molality, unlike molarity, does not vary with temperature.

Part F: Preparation of a Standard KIO_3 Solution*

Clean and rinse your 100 mL volumetric flask with deionized water. *Carefully* weigh out between 0.20 and 0.30 g of KIO_3 onto a piece of weighing paper. (*When given a range of measurements, aim for the minimum.*) Record the mass and carefully transfer the sample to your volumetric flask. (It is okay for the flask to be wet with deionized water.) Add about 50 mL of deionized water to the flask, and swirl carefully until all solid is dissolved. Then add enough water to bring the solution to the graduation mark. (If you overshoot, you must start over...*the results of your next lab depend on this solution!!*)

Cover the flask securely with its cap or a piece of parafilm, and then invert it 25–30 times to mix it thoroughly. Rinse the 250 mL plastic bottle provided by your instructor with deionized water then with a few milliliters of this solution; discard the waste. Pour the remaining solution into the bottle and label the bottle with the molarity, the date, your name, and lab section. Do not forget to cap the bottle tightly and give it to your instructor as you leave.

*Normally solid standards are dried to remove water before solution preparation. The error is too small for the precision of the balances available to you, so this step will be omitted.

Safety Notes

You will be using solutions of table salt and sugar among others. The table salt and sugar have been exposed to the laboratory environment and are no longer suitable for food.

17.3 WASTE DISPOSAL

The solution for Part F will be saved for use next week. All other solutions may be flushed down the drain with plenty of tap water.

EXPERIMENT 17: Concentration Units

DATA SHEET 1

A. Preparation of Molar Solution

Assigned Molarity _____

Volume of Solution _____

Mass of Sucrose _____

Instructor's initials:

B. Dilution of "Stock Solutions"

Assigned Final Volume _____

Molarity of "Stock Solution" _____

Volume of "Stock Solution" _____

Molarity of Final Solution _____

C. Molarity of Ionic Solutions

Assigned Mass _____

Assigned Ion _____

Volume of Solution _____

Molarity of Assigned Ion _____

Instructor's initials:

EXPERIMENT 17: Concentration Units

DATA SHEET 2

D. Percent by Mass of Solution

Assigned Mass of Solution _____

Assigned % (m/m) _____

Mass of Sucrose _____

Mass of Water _____

E. Preparation of a Molal Solution

Assigned Mass of Water _____

Assigned Molality _____

Mass of Sucrose _____

Mass of Prepared Solution _____

F. Preparation of a Standard KIO_3 Solution

Volume of Solution _____

Mass of KIO_3 _____

Molarity of KIO_3 _____

Instructor's initials:

EXPERIMENT 17: Concentration Units

DATA SHEET 3

A. Preparation of Molar Solution

B. Dilution of "Stock Solution"

C. Molarity of Ionic Species

EXPERIMENT 17: Concentration Units

DATA SHEET 4

D. Percent by Mass of Solution

E. Preparation of a Molal Solution

F. Preparation of Standard KIO$_3$ Solution

EXPERIMENT 17: Concentration Units

POSTLAB EXERCISE

1. Borax $(Na_2B_4O_7 \cdot 10H_2O)$ is a natural mineral that forms a basic solution in water. It is added to many detergents as a water softener. If _____ g of Borax is dissolved in water and the resulting solution is diluted to _____ mL, what is the molarity of Borax in this solution?

2. If the solution that you prepared in step B was placed inside an oven at 85 °C for 20 minutes, which would change: its molarity, molality, or percent by mass? Assume no evaporation. Explain your reasoning.

3. What mass of Na_3PO_4 is needed to prepare _____ mL of a solution that is _____ M PO_4^{3-}?

4. What would the molarity of Na^+ be in the solution prepared in question 3?

EXPERIMENT 18: Determination of Bleach: A Redox Titration

PRELAB EXERCISE

Terms:

Titration –

End point –

Equivalence point –

Redox reaction –

% (m/m) –

Safety Warnings:

EXPERIMENT 18: Determination of Bleach: A Redox Titration

PRELAB EXERCISE

1. To standardize a thiosulfate ($S_2O_3^{2-}$) solution, 15.00 mL of a 0.0419 M IO_3^- *(aq)* solution are pipetted into an 250 mL Erlenmeyer flask and treated with excess I^- to generate I_3^- *(aq)* according to Equation 3. How many millimoles of I_3^- *(aq)* are generated?

2. 20.08 mL of thiosulfate ($S_2O_3^{2-}$) solution are required to titrate the solution in question 1. The reaction taking place is represented by Equation 2. What is the molarity of thiosulfate ($S_2O_3^{2-}$) in the solution?

Use the following information to answer the questions below:

2.50 g of a bleach solution are placed into an Erlenmeyer flask and treated as described in Part B of your lab. The standardized $S_2O_3^{2-}$ solution from question 2 is used in the titration and 20.61 mL are required to reach the equivalence point.

3. How many millimoles of $S_2O_3^{2-}$ are required to reach the end point?

4. What are the corresponding millimoles of I_3^-?

5. What are the corresponding millimoles of ClO^-?

6. What is the percent by mass of NaClO in bleach solution?

18

DETERMINATION OF BLEACH: A REDOX TITRATION

Objectives

1. To gain experience with redox titrations.

2. To gain experience working with stoichiometric ratios.

3. To determine the concentration of sodium hypochlorite (NaClO) in a commercial bleach solution.

18.1 INTRODUCTION

Bleach (the common clothes brightener, toilet whitener, germ terminator) comes in two forms: liquid and powder. The active ingredient in the liquid bleach is NaClO, while the active ingredient in the powdered form of bleach is $Ca(ClO)_2$. Bleaches are also commonly known by the trade names Clorox, Purex, among others. An important question to ask is, "Which bleach has the greatest percent active ingredient, that is, the hypochlorite ion (ClO^-)?" In this experiment the percent of hypochlorite ion in various brands of bleach will be determined. The assumption is that the brand of bleach with the largest percentage of active ingredient is the most effective bleach.

Determination of Hypochlorite

In this experiment, the hypochlorite content of liquid bleach will be determined using a REDOX titration. First, iodide (I^-) in the form of KI is added in excess to a prepared sample of bleach:

$$ClO^-(aq) + 3I^-(aq) + H_2O(l) \rightarrow I_3^-(aq) + Cl^-(aq) + 2OH^-(aq) \qquad \text{(Equation 1)}$$

Then the resultant solution is acidified, and the triiodide (I_3^-), which formed as a product, is titrated with a standardized thiosulfate ($S_2O_3^{2-}$) solution until the solution is a *faint yellow* (a very small amount of I_3^- remains):

$$I_3^-(aq) + 2S_2O_3^{2-}(aq) \rightarrow 3I^-(aq) + S_4O_6^{2-}(aq) \qquad \text{(Equation 2)}$$

At this point, a soluble starch indicator is added, giving a deep blue color (the infamous "starch test" from biology); further dropwise addition of $S_2O_3^{2-}(aq)$ will cause the blue color to disappear. This marks the *equivalence point* of the titration.

So what did you do? From **Equation 1**, 1 mole of $ClO^-(aq)$ reacting with excess $I^-(aq)$ would produce 1 mole of $I_3^-(aq)$. From **Equation 2**, 1 mole of $I_3^-(aq)$ would react with 2 moles of $S_2O_3^{2-}(aq)$ to reach the equivalence point. So what is the mole relationship between $S_2O_3^{2-}(aq)$ and $ClO^-(aq)$?

Standardization of Thiosulfate

It is necessary to have a standardized thiosulfate solution (from $Na_2S_2O_3$) for the titration above. Thiosulfate can be standardized by titration of a known amount of $I_3^-(aq)$ solution (previously prepared in Experiment 17) with excess $I^-(aq)$ (from KI):

$$IO_3^-(aq) + 8I^-(aq) + 6H^+(aq) \rightarrow 3I_3^-(aq) + 3H_2O(l) \qquad \text{(Equation 3)}$$

Note: This titration will also employ the blue starch complex as an indicator.

18.2 EXPERIMENTAL

Part A: Standardization of 0.1XX M $Na_2S_2O_3$ Solution

Pipet 10.00 mL of 0.01X M KIO_3 into a 250 mL Erlenmeyer flask. (Remember the KIO_3 solution was previously prepared in Experiment 17. Be sure and record the molarity of KIO_3 calculated in Experiment 17.)

Add about 45 mL of deionized water, about 2 g of KI, and 10.0 mL of 0.5 M H_2SO_4. The 2 g of KI is an excess amount so exactly 2 g is not critical; however, do not be wasteful! Use a small amount of deionized water to wash any chemicals from the sides of the flask into the bottom of the flask to be reacted. Obtain a magnetic stirrer and a magnetic stir bar. Clean the magnetic stir bar, place it in the flask, and set the flask on a magnetic stirrer. Turn on the magnetic stirrer so that the stir bar is gently moving. **Do not** turn on the heating component of the magnetic stirrer. The magnetic stirrer will mix the components of the flask.

Properly clean and fill a 50 mL burette with the unstandardized 0.1XX M $Na_2S_2O_3$. Make sure that the burette tip is free of air bubbles and record the initial volume to the nearest 0.01 mL.

When the solution in the Erlenmeyer flask is well mixed and all solids have been dissolved, titrate the solution in the flask with your $S_2O_3^{2-}$ solution in the burette until the red-brown solution (excess I_3^-) changes to a yellow color. Add 2 drops of your starch indicator, and titrate dropwise, swirling constantly until the blue color disappears. Record the volume from your burette at the endpoint. (*Note: The change from blue to clear is the endpoint, <u>not the color change from red-brown to yellow!</u>*)

Repeat this procedure until you have performed a total of three titrations to standardize the $Na_2S_2O_3$ solution.

Part B: Preparation and Titration of Liquid Bleach

Weigh 2.00 to 3.00 g of bleach (approximately 45 drops) into a 250 mL Erlenmeyer flask and add about 45 mL of deionized water, 2 g of KI, and 10.0 mL of 3 M H_2SO_4. (*When given a range of measurements, aim for the minimum.*) A darkish color will develop indicating the formation of I_3^-.

Important:

1. *The chemicals added to the bleach should be added in the order listed!*

2. *Add the acid last and slowly as Cl_2 may be given off if these instructions are not followed.*

3. *The amount of KI needs to be measured accurately in this step.*

4. *Unless the same mass of bleach is obtained, remember that the volume of titrant required may differ significantly.*

Refill your burette with your now-standardized $Na_2S_2O_3$. Take an initial volume reading and titrate the solution in your Erlenmeyer flask until the darkish color is almost gone, or becomes a pale orange. (This is <u>not</u> the endpoint so *do not record* this as the final volume.) Add the starch indicator to the Erlenmeyer flask and titrate dropwise until the blue color disappears. (This <u>is</u> the endpoint. *Record* the final burette volume.)

Repeat the titration with additional samples of the bleach until you have performed a total of three titrations in this part. Be sure to use the same sample and to measure at the same balance.

Reminder: The titrant volume between trials will only be very close to each other if you weighed exactly the same mass of bleach for each titration! Your calculated percent by mass NaClO in the bleach should be fairly close, however.

Calculation for percent by mass active ingredient in bleach:

$$\% \text{ (m/m) NaClO} = \frac{\text{mg of NaClO}}{\text{mg of bleach}} \times 100\%$$

Safety Notes

Sulfuric acid is a strong acid. Chemicals must be added in the order directed.

18.3 Waste Disposal

All solutions may be flushed down the drain with plenty of tap water.

EXPERIMENT 18: Determination of Bleach: A Redox Titration

DATA SHEET 1

Part A: Standardization of 0.1XX M Na$_2$S$_2$O$_3$

	Trial 1	Trial 2	Trial 3
Experimental:			
Concentration of KIO$_3$ (from Experiment 17)	_____	_____	_____
Volume of KIO$_3$ Pipetted	_____	_____	_____
Final Burette Reading	_____	_____	_____
Initial Burette Reading	_____	_____	_____
Calculations:			
mmol of KIO$_3$ (IO$_3^-$)	_____	_____	_____
mmol of I$_3^-$ Formed (from Equation 3)	_____	_____	_____
Volume of Na$_2$S$_2$O$_3$ *(aq)* Required	_____	_____	_____
mmol of S$_2$O$_3^{2-}$ Present (from Equation 2)	_____	_____	_____
Molarity S$_2$O$_3^{2-}$	_____	_____	_____
Avg. Molarity Na$_2$S$_2$O$_3$		_____	

EXPERIMENT 18: Determination of Bleach: A Redox Titration

DATA SHEET 2

Part B: Preparation and Titration of Diluted Bleach

Experimental:

Average Molarity $Na_2S_2O_3$ _____

Brand of Bleach _____

	Trial 1	Trial 2	Trial 3
Mass of Bleach	_____	_____	_____
Final Burette Reading	_____	_____	_____
Initial Burette Reading	_____	_____	_____

Calculations:

	Trial 1	Trial 2	Trial 3
Volume of $Na_2S_2O_3$ *(aq)* Required	_____	_____	_____
mmol of $S_2O_3^{2-}$ Required	_____	_____	_____
mmol of I_3^- Present (Equation 2)	_____	_____	_____
mmol of ClO^- Present (Equation 1)	_____	_____	_____
mmol NaClO Present	_____	_____	_____
Mass of NaClO Present (mg)	_____	_____	_____
% (m/m) of NaClO in Bleach	_____	_____	_____
Avg. % (m/m) of NaClO in Bleach	_____		

EXPERIMENT 18: Determination of Bleach: A Redox Titration

POSTLAB EXERCISE

1. To standardize a thiosulfate ($S_2O_3^{2-}$) solution, _____ mL of a _____ M IO_3^- *(aq)* solution are pipetted into a 250 mL Erlenmeyer flask and treated with excess I^- to generate I_3^- *(aq)*. If _____ mL of the thiosulfate ($S_2O_3^{2-}$) solution are required to reach the equivalence point, what is the molarity of thiosulfate ($S_2O_3^{2-}$) in the solution?

2. _____ g of a bleach solution are placed into an Erlenmeyer flask and treated as described in Part B of your lab. The titration with _____ M $S_2O_3^{2-}$ solution (from question 1) requires _____ mL to reach the equivalence point. What is the percent by mass of NaClO in the bleach solution?

EXPERIMENT 19: Freezing Point Depressions

PRELAB EXERCISE

Terms:

     % (m/m) –

     Molality –

     Colligative properties –

Safety Warnings:

EXPERIMENT 19: Freezing Point Depressions

PRELAB EXERCISE

1. A graph of freezing point depression versus concentration is generated using data obtained from determining the ΔT_f of five known molal solutions according to the directions in today's write-up. What information from the graph is used to find your experimentally determined K_f for water?

    A solvent, S, having a density of 1.05 g/mL and $K_f = 3.90$ °C/m was uniformly cooled. A graph of temperature-time readings showed a plateau (flat) region [S(l)/S(s)] at 6.15 °C. A solution of 2.74 g compound B was dissolved in 15.5 mL of S. The solution was uniformly cooled and dropped in temperature until reaching 3.97 °C. It remained at that temperature for several minutes.

2. Determine the freezing point of solvent S.

3. How many grams of S were used?

4. What was the molality of compound B in the solution?

5. What is the molecular weight of compound B?

19

FREEZING POINT DEPRESSIONS

Objectives

1. To learn about colligative properties.

2. To determine the freezing point depression constant for water.

19.1 INTRODUCTION

A solution is a homogeneous mixture of two or more components. For a common two-component solution, the substance present in the major proportion is called the solvent and that in the minor proportion is called the solute. When solutes are dissolved in a liquid solvent, the vapor pressure of the latter is reduced; and, in general, the liquid range is extended to lower and to higher temperatures—that is, the solution freezes at a lower temperature than does the pure solvent and it boils at a higher temperature. These are examples of colligative properties. The magnitude of the vapor-pressure lowering (ΔV_p), the freezing-point depression (ΔT_f), and the boiling-point elevation (ΔT_b) are found experimentally to depend upon the number of solute particles and the nature of the solvent. Therefore, experimental measurement of ΔV_p, ΔT_f, and ΔT_b allows one to count the number of solute particles and thus calculate the molar quantity of solute in a given quantity of solvent. Or conversely, one can alter the properties of a solvent by systematically adding measured amounts of solute.

In this experiment you will be investigating the linear relationship between m_{solute} and ΔT_f for water. Given the equation:

$$\Delta T_f = m_{solute} (K_f) \qquad \text{or} \qquad K_f = \Delta T_f / m_{solute} \qquad \text{(Equation 1)}$$

where $\Delta T_f = T_f \text{(solvent)} - T_f \text{(solution)}$. Note: This definition of the Δ term is used for historical purposes to give a positive value of K_f and refers to freezing point depression, not change in freezing point.

The slope of the line should be equal to the K_f (freezing point depression constant for water). It should be noted that K_f is solvent specific—that is, different solvents have different values for K_f. For example, for p-xylene, C_8H_{10}, $K_f = 4.30$ °C/m and for pure acetic acid, CH_3COOH, $K_f = 3.90$ °C/m. The term m_{solute} is derived from the total number of solute particles and is independent of the type of particles.

For this experiment you will first determine the cooling curve and the freezing point for water dependent upon different amounts of solute. The second determination will be that of the value for the K_f of water. The class's collected experimental data for the freezing point depressions will be plotted on a graph versus the m_{solute}, and the slope of the line will be the value for K_f. In the postlab, this experimentally determined K_f of water will be compared to the accepted K_f of water which is 1.86 °C/m.

19.2 Experimental (work in pairs)

Part A

Obtain the following items: two thermometers, stopwatch, large test tube, rubber stopper with two holes to fit the test tube, copper stirring rod, salt, and ice.

Preparation of ice bath:

Obtain two 400 mL beakers of ice. Weigh out approximately 300 g of ice. In a third beaker, weigh out approximately 100 g of salt. In a large beaker (600 or 800 mL) layer the ice and salt beginning with the ice. Throughout the experiment additional ice and salt layers may be added to the ice bath. It is not necessary to weigh the additional salt and ice. The above measurements are to give an approximate idea of the proper ice–salt mixture. After the large test tube is prepared with the appropriate mixture of solute and solvent, insert the test tube into the center of the ice bath.

Preparation of test tube:

If not already assembled, carefully place the thermometer in one of the holes of the stopper, and place the stirring rod in the other hole. (Probably it will be necessary to lubricate the thermometer with glycerin before attempting to insert the thermometer into the rubber stopper. Be sure to clean the excess glycerin off the thermometer apparatus after the thermometer apparatus is inserted into the rubber stopper as the excess glycerin will affect your results.) Place the loop of the stirring rod around the thermometer. The tip of the thermometer and stirring rod should be located near the bottom of the test tube when the rubber stopper is placed into the test tube.

Freezing point determinations of solute/solvent mixtures:

Prepare a 30% (m/m) solution by mixing 9.00 g of glycerin with 21.00 g of water. (The total amount of solution for each of your mixtures is 30.00 g.) Tare a 50 mL beaker and weigh 21.00 g of water in the beaker. Add 9.00 g of glycerin to the water. (You may want to use a dropper as you approach the 30.00 g limit.) Stir the solution carefully until all of the glycerin appears to be dissolved in the water. Pour the solution into the test tube and place the rubber stopper on the test tube. Record the initial temperature of the solution in the test tube. Before determining the freezing point, use a second thermometer to make sure the ice bath temperature is at least -15 °C. If it is not, ask your lab instructor for additional instructions. Place the test tube in the ice bath. As soon as the test tube is in the ice bath, start the stopwatch. Record the temperature of the solution to the nearest 0.1 °C every 30 seconds. Continue recording the temperature until six consecutive readings are the same. (Since you are working in pairs for this experiment, it is recommended that one partner be responsible for the time increments while the other partner is responsible for stirring the solution in the large test tube. This will help maintain consistency in technique between trials.) Continuously stir the solution in the test tube to prevent super-cooling. The stirring is achieved by moving the stirring rod up and down throughout the solution at a fairly rapid pace. When you have completed the determination, be sure to thoroughly rinse the test tube with deionized water then clean and dry the thermometer apparatus before beginning another trial. Repeat the freezing point determination as described above to determine the freezing points for 25, 20, 15, and 10% (m/m) mixtures.

Freezing point determination of pure solvent:

Measure 30 g of deionized water in a tared beaker. Pour the water into the large test tube and place the rubber stopper on the test tube. Record the initial temperature of the water in the test tube and place the test tube in the ice bath. Complete the experimental portion of this lab by repeating the freezing point determination as described above.

Part B

Each group will provide their ΔT_f information for this part (see Data Sheet 3). Referring to "Appendix E: Generating a Straight-Line Graph in Microsoft Excel", plot a linear graph on the computer using the collected experimental data for ΔT_f (y-axis) versus the molality of the solute (x-axis). Convert percent by mass into molality using the molecular weight of the solute. Be sure to turn in your graph with your report.

Example Molality Calculation:

Given a 10% (m/m) solution prepared by mixing 3.00 g of glycerin with 27.00 g of water, the molality would be calculated as follows:

$$\text{(3.00 g glycerin/0.027 kg water)(1 mole glycerin/92.09 g glycerin)}$$
$$= 1.21 \ m_{glycerin}$$

From Equation 1: ΔT_f is the y-axis and m_{solute} is the x-axis. The slope of the line is the K_f value for water. Record this on your data sheet to be turned in along with your graph.

Safety Notes

Be careful placing the rubber stopper on the test tube as the test tube could break. Let the solid melt before removing the thermometer and stirring rod as the thermometer could break.

19.3 Waste Disposal

All solutions may be flushed down the sink with tap water.

19.4 Graphing

Reference "Appendix E: Generating a Straight-Line Graph in Microsoft Excel" for instructions on generating the graph for this lab.

Construct a graph of the class average freezing point depressions versus the molality of the glycerin solutions. You are required to use the computer to generate this graph. Your lab instructor will help you with the graph during lab. You can also get help during your lab instructor's office hours.

EXPERIMENT 19: Freezing Point Depressions

DATA SHEET 1 – INDIVIDUAL DATA

% (m/m) Glycerin Solution	30%	25%	20%	15%	10%	0%
Experimental Mass of Glycerin	_____	_____	_____	_____	_____	_____
Experimental Mass of Water	_____	_____	_____	_____	_____	_____
Molality *(x-axis values on the graph)*	_____	_____	_____	_____	_____	_____

Note: The 0% is the pure solvent and is not included on the graph.

EXPERIMENT 19: Freezing Point Depressions

DATA SHEET 2 – TIME SHEET

% (m/m) Glycerin Solution Temperature (°C) at time	30%	25%	20%	15%	10%	0%
0:00	___	___	___	___	___	___
0:30	___	___	___	___	___	___
1:00	___	___	___	___	___	___
1:30	___	___	___	___	___	___
2:00	___	___	___	___	___	___
2:30	___	___	___	___	___	___
3:00	___	___	___	___	___	___
3:30	___	___	___	___	___	___
4:00	___	___	___	___	___	___
4:30	___	___	___	___	___	___
5:00	___	___	___	___	___	___
5:30	___	___	___	___	___	___
6:00	___	___	___	___	___	___
6:30	___	___	___	___	___	___
7:00	___	___	___	___	___	___
7:30	___	___	___	___	___	___
8:00	___	___	___	___	___	___
8:30	___	___	___	___	___	___
9:00	___	___	___	___	___	___
9:30	___	___	___	___	___	___
10:00	___	___	___	___	___	___
10:30	___	___	___	___	___	___
11:00	___	___	___	___	___	___
11:30	___	___	___	___	___	___
12:00	___	___	___	___	___	___
12:30	___	___	___	___	___	___
13:00	___	___	___	___	___	___
13:30	___	___	___	___	___	___
14:00	___	___	___	___	___	___
T_f of Water						___
T_f of % (m/m)	___	___	___	___	___	
$\Delta T_f = T_{f\ of\ water} - T_{f\ of\ mixture}$	___	___	___	___	___	

EXPERIMENT 19: Freezing Point Depressions

DATA SHEET 3 – GROUP DATA

ΔT_f **for:**

Group	30%	25%	20%	15%	10%
_____	_____	_____	_____	_____	_____
_____	_____	_____	_____	_____	_____
_____	_____	_____	_____	_____	_____
_____	_____	_____	_____	_____	_____
_____	_____	_____	_____	_____	_____
_____	_____	_____	_____	_____	_____
_____	_____	_____	_____	_____	_____
_____	_____	_____	_____	_____	_____
_____	_____	_____	_____	_____	_____
_____	_____	_____	_____	_____	_____
_____	_____	_____	_____	_____	_____
_____	_____	_____	_____	_____	_____
_____	_____	_____	_____	_____	_____
_____	_____	_____	_____	_____	_____

Average ΔT_f   _____   _____   _____   _____   _____
(y-axis values on the graph)

EXPERIMENT 19: Freezing Point Depressions

POSTLAB EXERCISE

1. What is your experimentally determined value for the K_f for water?

2. The accepted K_f for water is 1.86 °C/m. What is the percent error for your value?

3. What do you think may cause this error?

4. Why is molality used in this experiment to express concentration instead of molarity?

EXPERIMENT 20: Chemical Kinetics: The Iodine Clock

PRELAB EXERCISE

Terms:

      Chemical kinetics –

      Rate of reaction –

      Rate law –

      Method of initial rates –

Safety Warnings:

EXPERIMENT 20: Chemical Kinetics: The Iodine Clock

PRELAB EXERCISE

1. What is the concentration of $S_2O_3^{2-}(aq)$ in a solution prepared by mixing 25.00 mL of 0.050 M $Na_2S_2O_3(aq)$ with enough 0.200 M KCl(aq) to make 100.0 mL of solution? (Assume volumes are additive and that there is no chemical reaction between the compounds involved.)

2. If all the $S_2O_3^{2-}(aq)$ in the solution above is completely reacted with another substance over a 35.0-second period, what is the rate of decrease of the $S_2O_3^{2-}(aq)$ in solution? (Assume no volume change occurs.)

Use the following data set to answer the questions below:

$$A(aq) + B(aq) \rightarrow \text{Products}$$

Rxn	[A]	[B]	Rate of Loss of A (M/s)
1	0.25	0.25	9.0×10^{-3}
2	0.25	0.50	3.6×10^{-2}
3	0.50	0.25	1.8×10^{-2}

3. Determine the orders of reaction (m and n) for reactants A and B. Explain how you came to this conclusion or show your work.

4. Determine the rate constant (k) for the reaction above.

20

CHEMICAL KINETICS: THE IODINE CLOCK

Objectives

1. To use the iodine clock reaction to study the kinetics of a redox reaction.

2. To determine the order of a chemical reaction.

3. To determine the rate law constant of a chemical reaction.

20.1 INTRODUCTION

The determination and study of rates of reactions is referred to as **chemical kinetics**. It is frequently very difficult to determine when a reaction has gone to completion, so the study of the kinetics of a reaction is usually concerned with initial rates—that is, the average rate of a reaction over the first few seconds, or at most a few minutes, during which time the reactant concentrations have varied only slightly from their original values.

The peroxodisulfate ion, $S_2O_8^{2-}$, oxidizes iodide ion reasonably slowly at room temperature in accord with the equation:

$$2\,I^-(aq) + S_2O_8^{2-}(aq) \rightarrow I_2(aq) + 2SO_4^{2-}(aq) \qquad \text{(Equation 1)}$$

The **rate of reaction** can be expressed simply in terms of the decrease of any one of the reactants with respect to time or the increase of one of the products with respect to time and can be written arithmetically:

$$\text{Rate} = -\frac{\Delta[S_2O_8^{2-}]}{\Delta t} = -\frac{[I^-]}{2\Delta t} = \frac{\Delta[I_2]}{\Delta t} = \frac{[SO_4^{2-}]}{2\Delta t} \qquad \text{(Equation 2)}$$

The first equality above states that the reaction rate is equal to the decrease in peroxodisulfate ion concentration for a given time interval. Since the stoichiometry (Equation 1) indicates that two iodide ions are used for every peroxodisulfate ion that reacts, the second equality, involving the iodide ion, shows this fact, and so forth.

One of the main purposes of a kinetics experiment is to find the **rate law** for the reaction—that is, the algebraic expression that relates rate to the concentration(s) of the reactant(s). In the above reaction the rate depends both on the concentration of I^- and $S_2O_8^{2-}$ ions and takes the general form:

$$\text{Rate} = k[I^-]^m[S_2O_8^{2-}]^n \qquad \text{(Equation 3)}$$

where k is the **rate constant** and m and n are the **reaction orders**.

In this experiment you will study the kinetics of the reaction of iodide ion with peroxodisulfate ion and determine the rate law. Our method for measuring the rate of reaction involves what is frequently called a "clock" reaction. This is a second reaction (or set of rapid consecutive reactions) that gives a signal, such as a color change, when a particular quantity of one of the reactants (here, I^- or $S_2O_8^{2-}$) has been consumed. Our "clock" will be based on the facts that (i) thiosulfate ion, $S_2O_3^{2-}$, reacts rapidly and quantitatively with iodine, I_2:

$$I_2(aq) + 2S_2O_3^{2-}(aq) \rightarrow 2I^-(aq) + S_4O_6^{2-}(aq) \qquad \text{(Equation 4)}$$

and (ii) an intense blue-black coloration is formed almost instantly when exceedingly small traces of I_2 are in the presence of starch and I^- (hence, an indicator that detects iodine). Thus, the reaction of peroxodisulfate ion with iodide ion is started in the presence of a known trace amount of thiosulfate ion and the starch indicator. As the reaction progresses, the iodine, as it is formed, immediately reacts with the thiosulfate and the reaction mixture remains colorless. When the trace quantity of thiosulfate is used up, iodine produced by the oxidation of iodide ion now remains in solution and causes the starch indicator to turn blue-black in color. Therefore, at time t, the time after mixing when the solution turns blue-black, for every mole of thiosulfate originally present, 1/2 mole of iodine has been produced and used up, and 1/2 mole of peroxodisulfate also has been used up. At this point, the change in concentration of $S_2O_8^{2-}$ must be equal to half the original $S_2O_3^{2-}$ concentration, and thus, where a rate has been expressed in terms of $-\Delta[S_2O_8^{2-}]/\Delta t$, or

$$\text{Rate} = -\frac{\Delta[S_2O_8^{2-}]}{\Delta t} = \frac{1}{2}\frac{[S_2O_3^{2-}]_{orig}}{\Delta t} \qquad \text{(Equation 5)}$$

The calculation of the **rate constant** (k) and the **orders of the reaction** (m and n) with respect to I^- and $S_2O_8^{2-}$ follow from the dependence of the reaction rate on reactant concentrations. (Compare Equations 3 and 5.)

20.2 EXPERIMENTAL (WORK IN PAIRS)

Uncatalyzed Reactions

The following preparation table shows the reaction mixtures 1–5 to be studied in order to determine the rate constant and reaction orders for the reaction of iodide ion with peroxodisulfate ion. **The KI and $(NH_4)_2S_2O_8$ solutions should be very carefully measured with appropriately sized graduated cylinders, and the $Na_2S_2O_3$ solution should be dispensed with a 10 mL volumetric pipet or a burette.** The KCl and $(NH_4)_2SO_4$ solutions may be measured out in graduated cylinders.

Table for the Preparation of Experimental Runs

Rxn	mL 0.200 M KI	mL 0.200 M KCl	mL 0.100 M $(NH_4)_2S_2O_8$	mL 0.100 M $(NH_4)_2SO_4$	mL 0.0050 M $Na_2S_2O_3$
1	20.0	0	20.0	0	10.00
2	10.0	10.0	20.0	0	10.00
3	20.0	0	10.0	10.0	10.00
4	20.0	0	5.0	15.0	10.00
5	15.0	5.0	15.0	5.0	10.00
1A	20.0	0	20.0	0	10.00

Carefully measure the KI using a graduated cylinder (I) and the $(NH_4)_2S_2O_8$ into a similar graduated cylinder (II) for Rxn 1. The 10.00 mL of $Na_2S_2O_3$ solution and 3–4 drops of starch solution are placed in a 250 mL Erlenmeyer flask (along with the required volumes of KCl and $(NH_4)_2SO_4$ solutions in Rxns 2–5). These last two components are added merely to dilute each reaction mixture to a total volume of 50.0 mL while keeping the ionic strength constant. Start the reaction by simultaneously pouring the KI and $(NH_4)_2S_2O_8$ solutions from graduated cylinders I and II into the Erlenmeyer flask and swirling thoroughly. At the instant the reactants are mixed, the stopwatch should be started. Stop timing when the first sudden appearance of the blue-black color occurs. Record to the nearest second the time required for the reaction. Also, record the final temperature of the reaction mixture. Repeat the procedure above for each of the four other reaction mixtures.

Catalyzed Reaction

Metallic cations have a pronounced effect on the rate of this reaction—that is, they may act as catalysts. You may observe this by carrying out Rxn 1A, which is a <u>duplicate of Rxn 1</u> except that you are to add 1 drop of 0.1 M $CuSO_4(aq)$ to the ammonium peroxodisulfate solution in graduated cylinder II. Stir graduated cylinder II for a few seconds to mix the catalyst and then mix the solutions together in the 250 mL Erlenmeyer flask to determine the time when the color change occurs. Compare this with Rxn 1.

Each group will provide their time data to be compared with the data from other groups.

20.3 Calculations

For the rate law

$$\text{Rate} = -\frac{\Delta[S_2O_8^{2-}]}{\Delta t} = k_T[I^-]^m[S_2O_8^{2-}]^n$$

(Equation 6)

use the **method of initial rates** (as described in class and below) to evaluate and record both the reaction "orders" (*m* and *n*) and evaluate and record the apparent rate constants

(k_T) from the data of Rxns 1–5. Report the average value of k_T and show its units. Also, comment on the results of the catalyzed Rxn 1A relative to Rxn 1.

Determination of Rate

Since the $[S_2O_3^{2-}]_{orig}$ is the same in each of the reactions, and since it is assumed to be completely used up by the time (t_b) the color change occurs, the following relationship can be shown to be true (using Equation 5):

$$\text{Rate of loss of reactant, } S_2O_8^{2-} = -\frac{1}{2}\frac{\Delta[S_2O_3^{2-}]}{\Delta t} = \frac{5.00 \times 10^{-4}\,M}{t_b}$$

where t_b = time required for the clock to "tick." The value 5.00×10^{-4} M is the same for each flask because it depends on the limiting reactant, the $S_2O_3^{2-}$.

Determination of Rate Law

If you have been careful in solution preparation, the **rate law** can be determined for your reaction:

$$\text{Rate} = -\frac{\Delta[S_2O_8^{2-}]}{\Delta t} = k_T[I^-]^m[S_2O_8^{2-}]^n$$

Where *m* and *n* are called the reaction orders for $[I^-]$ and $[S_2O_8^{2-}]$, respectively.

Determination of *m* and *n* is made by inspection (more accurately by mathematics) using the method of initial rates.

Method of Initial Rates

The **method of initial rates** measures the rate of a reaction just after the reactants are brought together and allowed to interact. To determine the orders of the reaction, individual reactions are compared to see how rate is affected by changing the initial concentrations of one reactant, while holding the concentration of the other reactant constant. Consider the following simplified possibilities:

If the $[A]_{init.}$ in Reaction 1 is doubled in Reaction 2 while $[B]_{init.}$ is the same in Reactions 1 and 2, the rate of the reaction may:

1. remain **unchanged**—if so the reaction order for [A] is **0.**

2. be **doubled**—if so the reaction order for [A] is **1.**

3. be **quadrupled**—if so the reaction order for [A] is **2.**

Thus, for the purpose of this experiment, the orders of reaction (*m* and *n*) can have values of 0, 1, or 2.

Example Data:

Rxn	[A]	[B]	Rate of Loss of A (M/s)
1	0.15	0.15	4.0×10^{-2}
2	0.15	0.30	4.0×10^{-2}
3	0.30	0.15	1.6×10^{-1}

A*(aq)* + B*(aq)* → Products

*Between Rxn 1 and Rxn 2, [A] is **constant** while [B] is **doubled** (i.e., 0.15 M → 0.30 M). The rate of the reaction is **unchanged**, so the order of reaction for reactant B is *n = 0*.

*Between Rxn 1 and Rxn 3, [B] is **constant** while [A] is **doubled** (i.e., 0.15 M → 0.30 M). The rate of the reaction is **quadrupled**, so the order of reaction for reactant A is *m = 2*.

The rate law is therefore: Rate of Loss of A = $k[A]^2 [B]^0$

or more simply Rate = $k[A]^2$.

The rate constant (k) can be solved for by algebraic substitution into the equation: $k = $ Rate/$[A]^2$.

For example: Using the data from Rxn 1 gives $k = 1.8$ $M^{-1}s^{-1}$. For a given reaction, at a given temperature, k should be a constant that is independent of each of the reactant concentrations.

Safety Notes

Some of these solutions may temporarily stain your skin on contact.

20.4 WASTE DISPOSAL

All solutions may be flushed down the sink with water.

EXPERIMENT 20: Chemical Kinetics: The Iodine Clock

DATA SHEET 1

Table 1: Preparation

Rxn	mL 0.200 M KI	mL 0.200 M KCl	mL 0.100 M $(NH_4)_2S_2O_8$	mL 0.100 M $(NH_4)_2SO_4$	mL 0.0050 M $Na_2S_2O_3$	Temp. in °C	t, Time Required in Seconds
1	20.0	0	20.0	0	10.00		
2	10.0	10.0	20.0	0	10.00		
3	20.0	0	10.0	10.0	10.00		
4	20.0	0	5.0	15.0	10.00		
5	15.0	5.0	15.0	5.0	10.00		
1A	20.0	0	20.0	0	10.00		

EXPERIMENT 20: Chemical Kinetics: The Iodine Clock

DATA SHEET 2

Group Data

Group Members	Rxn 1 Time (s)	Rxn 2 Time (s)	Rxn 3 Time (s)	Rxn 4 Time (s)	Rxn 5 Time (s)
Class Average					

EXPERIMENT 20: Chemical Kinetics: The Iodine Clock

DATA SHEET 3

Calculation Results (Use Class Average):

Table 3: Initial Rates

Rxn	$[I^-]$	$[S_2O_8{}^{2-}]$	Rate	Rate Constant, k
1				
2				
3				
4				
5				

Average value for k = _____

m = _____

n = _____

Rate Law: **Rate** = _____

EXPERIMENT 20: Chemical Kinetics: The Iodine Clock

POSTLAB EXERCISE

1. Referring to your experimental results, does the catalyst in Reaction 1A cause the rate of the reaction to increase, decrease, or stay the same as compared to Reaction 1? How do you know?

2. Based on your experimentally determined rate law, predict a value for <u>rate of reaction</u> if $[I^-] = [S_2O_8^{2-}] = 0.750$ M.

Use the following data set to answer the questions below:

$$A(aq) + B(aq) \rightarrow Products$$

Rxn	[A]	[B]	Rate of Loss of A (M/s)
1	0.12	0.10	8.4×10^{-6}
2	0.12	0.20	1.7×10^{-5}
3	0.24	0.10	3.4×10^{-5}

3. Determine the orders of reaction (m and n) for reactants A and B. Explain how you came to this conclusion or show your work.

4. Determine the rate constant (k) for the reaction above.

EXPERIMENT 21: Spectroscopy of Dyes

PRELAB EXERCISE

Terms:

Spectroscopy –

Absorbance –

Transmittance –

Beer's law –

Standard curve –

Safety Warnings:

EXPERIMENT 21: Spectroscopy of Dyes

PRELAB EXERCISE

1. Identify the four variables in the Beer's law equation.

2. A sample gives a percent transmittance of 63. What is its absorbance?

3. The sample in question 2 has a concentration of 0.121 M. The sample has a pathlength of 2.00 cm. What is the absorptivity of the sample?

4. If the sample in question 3 is diluted to a concentration of 0.0121 M, what will be the new absorbance? Note: Pathlength and absorptivity are constant.

21

SPECTROSCOPY OF DYES

Objectives

1. To gain experience using a simple spectrometer.

2. To determine the concentration of an unknown through a set of calibration standards.

3. To gain experience graphing a standard curve.

21.1 INTRODUCTION

Spectroscopy studies the interaction of light with matter. Colored analytes (species being studied) absorb light at specific wavelengths. When light of an appropriate wavelength passes through a sample, the emerging light is of lower intensity than that entering the sample (see Figure 1).

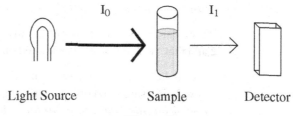

Light Source   Sample   Detector

Figure 1

where I_0 is the intensity of the incident light and I_1 is the intensity of the emerging light.

The ratio of I_1/I_0 is called the transmittance. The absorbance (A) is equal to the negative log of the ratio (I_1/I_0) or $\{A = -\log(I_1/I_0)\}$. The absorbance is proportional to three factors:

1. absorptivity of the sample

2. distance the light travels through the sample

3. concentration of the absorbing species

The relationship, expressed as **A = abc**, is known as Beer's law.

The SpectroVis Plus is an instrument that measures the percent transmittance of light at various wavelengths. The absorbance is found using $A = -\log(\%T/100)$. As can be seen from Beer's law there is a direct relationship between the absorbance, A, and the concentration, c, of the sample. If the values of the constants a and b are known, the concentration of a solution may be determined from its absorbance of light. Typically, a series of solutions of known concentrations is prepared and the absorbance of each of these solutions is measured. A standard curve of absorbance versus concentration should give a straight line with a slope of ab. Once the data are plotted, the graph can be used to determine the concentration of a solution for which the absorbance has been measured. This is done using a best-fit line.

21.2 EXPERIMENTAL (WORK IN PAIRS)

Preparation of the Standard Solutions

Obtain

- two clean, dry **100 mL** beakers and label them as **1** and **11**.
- nine clean, dry **50 mL** beakers and label them **2** to **10**.

In **Beaker 1**, obtain approximately 80 mL of **concentrated known dye solution**. Record the name of the dye, its concentration, and the required wavelength (λ max) in Data Sheet 2. Rinse a clean 50 mL burette with a small amount of the dye solution. Fill the burette to between 0 and 1 mL with the dye solution. Check for and remove any air pockets just under the stopcock before recording the initial volume.

In **Beaker 11**, obtain about 80 mL of **deionized water** (DI H_2O). Rinse a second clean 50 mL burette with a small amount of the DI H_2O then fill the burette with the DI H_2O. Check for and remove any air pockets just under the stopcock. Record the initial volume.

Begin warming up the SpectroVis Plus by making sure it is connected to the LabQuest 2 and that the LabQuest 2 is powered on.

Now, make your standard solutions. Carefully fill beakers 2–10 according to the table below, being sure to record your initial and final burette volumes in Data Sheet 1. *Note: Never drain the liquid in a burette below the 50 mL line. If you get close, ask your instructor for help.* These solutions, with your known dye solution in Beaker 1, are the 10 standards which will be used to generate a standard curve, and they *must be prepared carefully*.

Beaker	mL of known dye solution	mL deionized water
2	9.00	1.00
3	8.00	2.00
4	7.00	3.00
5	6.00	4.00
6	5.00	5.00
7	4.00	6.00
8	3.00	7.00
9	2.00	8.00
10	1.00	9.00

Obtain 12 clean cuvettes. Label each cuvette 1–12 at the top (or otherwise keep track of which is which). Fill cuvette 1 approximately ¾-full with the solution from beaker 1, which is the stock solution and your first standard. Fill cuvette 2 approximately ¾-full with the solution from beaker 2. Continue in this manner until cuvettes 3–10 are filled. Fill cuvette 11 with deionized water—this is your blank and will be used to calibrate the spectrophotometer. Fill cuvette 12 with the **UNknown** dye solution. Be sure the unknown is the same color as the stock solution used to make the standards. Record the unknown identity in Data Sheet 2.

Calibration and Setup of the Vernier SpectroVis Plus

The LabQuest 2 is an interface that operates using a touch screen. To navigate the interface, use the stylus attached to the right side of the instrument. When not in use, return the stylus to its storage slot. Before beginning the calibration, have all cuvettes ready for measuring.

1. The SpectroVis Plus (**SVP**) spectrophotometer should be connected to the LabQuest 2 (**LQ2**), and the LQ2 plugged in and powered on.

2. In the *uppermost left* corner of the LQ2 screen, the **Meter icon** (See Figure 1) should be the only icon displayed. NOTE: If it is not, tap the **Home icon** at the bottom right side of the screen then tap the **Meter icon**. Now, this should be the *only* icon in the uppermost left corner of the screen.

Figure 1: The Meter icon

3. Tap the word **File** then select **New** in the File menu. If prompted, choose **Discard of any previous data**. Tap the word **Sensors** then **Change Units** ▶ **USB: Spectrometer** ▶ **% Transmittance**.

4. To enter λ max into the LQ2, tap somewhere in the red box on the screen and select **Change Wavelength**. Enter λ max in nm (given on the bottle at the main bench) and select **OK**. The LQ2 may slightly adjust your wavelength; this is fine. For example, you enter 400 nm but the wavelength shown on the LQ2 is 399.7 nm.

5. In the **Sensors** menu (or tap somewhere in the red box on the screen), choose **Calibrate** ▶ **USB: Spectrometer**. The following message is displayed: "Taking dark sample". Select **OK**. The following message is displayed: "Waiting 90 seconds for lamp to warm up ...". If the lamp has already been warmed up at least 5 minutes, tap **Skip Warmup**. Otherwise, after 90 seconds, the message will change to "Warmup complete".

6. Prepare cuvette 11, the blank, as follows:

    -   Wipe the outside with a lint-free tissue.
    -   After wiping, handle it by the top edge only.
    -   Dislodge any bubbles by gently tapping it on the bench top.
    -   Position it in the SVP so the light passes through the clear sides. If your cuvette has two clear sides and two cloudy sides, make sure the clear sides are in the path of the light (*left to right*, *NOT top to bottom*).

7. When it becomes highlighted, select **Finish Calibration**. When the message "Calibration completed." appears on the screen, select **OK**.

   NOTE: If all 12 of your cuvettes are prepared properly and ready for measurement BEFORE you measure the blank (cuvette 11), you should only have to measure the blank once.

8. On the Meter screen, tap **Mode**. In the **Mode** menu select **Events with Entry**. Select **OK**. Enter the wavelength now displayed on the Meter screen. For water, the number shown should be close or equal to 100%, for example, 99.6%.

Measuring the Percent Transmittance of the Standard Solutions

Place cuvette 1, your most concentrated standard sample, in the spectrophotometer. In Data Sheet 2, record the percent transmittance shown on the LQ2 screen. Repeat with your remaining standards and your unknown. **If the unknown gives a reading lower than your third standard solution, dilute it in a known manner and measure again.** Account for this dilution in your data sheet.

Safety Notes

There are only generic safety concerns for this experiment.

21.3 WASTE DISPOSAL

All solutions may be flushed down the drain with plenty of tap water.

21.4 GRAPHING

Reference "Appendix E: Generating a Straight Line Graph in Microsoft Excel" for instructions on generating the graph for this lab.

Construct a graph of absorbance of the standard solutions versus concentration of the standard solutions. You are required to use the computer to generate this graph. Your lab instructor will help you with the graph during lab. You can also get help during your lab instructor's office hours.

EXPERIMENT 21: Spectroscopy of Dyes

DATA SHEET 1

Stock Solution

Dye color: _____   Dye concentration: _____

Required wavelength: _____

Dye Volumes (mL)

Soln	2	3	4	5	6	7	8	9	10
Final Vol.	___	___	___	___	___	___	___	___	___
Initial Vol.	___	___	___	___	___	___	___	___	___

Deionized Water Volumes (mL)

Soln	2	3	4	5	6	7	8	9	10
Final Vol.	___	___	___	___	___	___	___	___	___
Initial Vol.	___	___	___	___	___	___	___	___	___

EXPERIMENT 21: Spectroscopy of Dyes

DATA SHEET 2

Solution	Vol Dye	Vol H$_2$O	*Vol Soln	Concentration	% T	Absorbance
1	NA	NA	NA			
2						
3						
4						
5						
6						
7						
8						
9						
10						

Unknown

Unknown ID: _____

If you diluted the unknown, describe how: _____

Unknown % Transmittance: _____   Unknown Absorbance: _____

* Volumes are not necessarily additive. We are assuming that the density of the solutions remains constant when diluting, which enables us to add them together.

EXPERIMENT 21: Spectroscopy of Dyes

POSTLAB EXERCISE

1. What is the concentration of your original unknown solution?

2. A UV/Vis spectrophotometer was used to measure the absorbance of a solution of hemoglobin bound to oxygen at 590 nm. The molar absorptivity was determined to be 8.57×10^4 $M^{-1}cm^{-1}$. The cuvette used had a 1.00 cm sample pathlength. The measured absorbance was _____. What was the concentration of hemoglobin in the solution?

3. What would be the absorbance of the solution in question 1 if the pathlength were 2.00 cm?

EXPERIMENT 22: Le Châtelier's Principle

PRELAB EXERCISE

Terms:

      Equilibria –

      Exothermic reaction –

      Endothermic reaction –

      Le Châtelier's principle –

Safety Warnings:

EXPERIMENT 22: Le Châtelier's Principle

PRELAB EXERCISE

1. Name the three types of equilibria that you will be examining today and give the general reactions for each.

2. From today's write-up, give a specific example reaction for each of the types of equilibria in question 1.

22 LE CHÂTELIER'S PRINCIPLE

Objectives

1. To use Le Châtelier's principle to explain the observations seen when a reaction at equilibrium is disturbed by adding or removing reactants or products.

2. To use Le Châtelier's principle to explain the observations seen when a reaction at equilibrium is disturbed by adding or removing heat energy.

22.1 Introduction

Le Châtelier's principle simply states that if we apply a stress to, or disturb, a system that is in equilibrium, it will respond in order to reestablish equilibrium. Disturbances include but are not limited to:

1. the addition or removal of products

2. the addition or removal of reactants

3. a change in temperature; heating or cooling

In this lab you will look at Le Châtelier's principle as applied to three different types of equilibria: solubility equilibria, acid–base equilibria, and complex equilibria.

22.2 Experimental

Part A: Solubility Equilibria

A solubility equilibrium can be described by:

$$Solute + Solvent \rightleftharpoons Solution$$

Using an ionic solid as the solute and water as the solvent, we can investigate the equilibrium of a saturated solution of NaCl. Saturated implies that there is as much solute as possible dissolved in the solvent at a given temperature.

The solubility of NaCl can be represented by the following equilibrium:

$$NaCl(s) \rightleftharpoons Na^+(aq) + Cl^-(aq)$$

Obtain two small test tubes. Label them tube 1 and tube 2. Place the test tubes in a 50 mL beaker for support. Add the following:

Test tube 1   6 drops NaCl   0 drops H_2O

Test tube 2   0 drops NaCl   6 drops H_2O

Carefully add 6 drops of concentrated (12 M) HCl to each test tube. Gently stir each solution. If no change is observed in tube 1, repeat this process. Record the total amount of concentrated HCl added and your observations.

Pour the contents of the test tubes into the 50 mL beaker; add a spatula tip-full of baking soda. It will begin to bubble as carbon dioxide is released and the acidic mixture is neutralized. When the bubbling stops, add another spatula tip-full of baking soda. Repeat this procedure until a small addition of baking soda no longer causes the mixture to bubble. The mixture will then be neutralized and can be rinsed down the drain.

CAUTION: Concentrated (12 M) HCl produces biting fumes of gaseous HCl. It also burns the skin and damages clothing. Keep your supply under the hood. Immediately flush your skin with water if any acid gets on your skin. Notify your instructor of any spills.

Part B: Acid–Base Equilibria (Indicators)

There is a large group of chemical substances, called acid–base indicators, which change color in solution when $[H^+]$ changes. The generalized symbol for an indicator is Hin (the acid form, because it contains an H^+ in its formula) or In^- (the base form, because it has one less H^+ than the acid form). The equilibrium can be described as below:

$$Hin(aq) \rightleftharpoons H^+(aq) + In^-(aq)$$

acid color   base color

Adding H^+ to a solution would shift this equilibrium which way? What effect will adding OH^- have even though it is not in the reaction?

You will be looking at several indicators in this experiment. You will determine the colors for the acid and base forms, as well as see the effects of adding OH^- and H^+.

Obtain a well plate (spot plate with multiple wells) and several acid–base indicators. Prepare two wells for each indicator—one for the acid color and one for the base color. To each well add 5 drops of water and 2 drops of indicator.

Now to one well add 2 drops of 1.0 M HCl. (This will give the acid color of the indicator.) To the other well add 2 drops of 1.0 M NaOH. (This will give the base color of the indicator.) Record these colors. Now add 6 drops of the 1.0 M NaOH to your acidic well and 6 drops of 1.0 M HCl to your basic well. Record these colors. How does the equilibrium shift?

Repeat the tests above with the other indicators. (**Note:** It is likely that you can place all indicators on the spot plate at the same time. Be careful to record which indicator is in which well.) The available indicators may differ from those listed on the data page. Please alter the data page if necessary.

Part C: Complex Equilibria (Instructor Demonstration)

Ions can combine with other ions (or molecules) to form new species in solution. These are called complex ions, or more generally, complex species. Consider the following equilibrium:

$$Co(H_2O)_6^{2+}(aq) + 4Cl^-(aq) \rightleftharpoons CoCl_4^{2-}(aq) + 6\,H_2O(l)$$

(rose–red)   (brilliant blue)

Many transition metals and their complexes are quite colorful. This equilibrium represents the formation of a tetrachloro complex of cobalt.

Obtain six test tubes. Place 20 drops of 0.30 M $Co(NO_3)_2$ into each of the test tubes. Then add to each tube 12 drops of deionized water, followed by 45 drops of concentrated (12 M) HCl (CAUTION!), or enough to make the mixture purple. Mix these solutions thoroughly with small stirring rods. Record the color produced in these tubes. The color should be a shade of purple and should be the same in each test tube. If a test tube contains a paler purple than the others, add 12 M HCl dropwise until the color is the same.

Now set up a boiling water bath in a 250 mL beaker and an ice/water slurry in another 250 mL beaker. Place the first tube in the hot water, the second in the ice water, and leave the third at room temperature as a control. After a few minutes, compare the colors and record your results. Test to see if the reaction is reversible by placing the hot test tube in the cold water and vice versa.

In the fourth test tube place 10 drops of concentrated HCl. To the fifth test tube, add a few drops of 0.2 M $AgNO_3$. To the sixth test tube double the volume with water. Stir the tubes and compare the colors.

Waste for this part of the lab is toxic and should be collected in a designated waste container.

Complete your data sheet by recording your observations.

Safety Notes

See the special warning on the previous page about concentrated HCl. Caution should also be used with the 1.0 M HCl and the 1.0 M NaOH.

22.3 Waste Disposal

The acidic solutions from Part A should be reacted with baking soda until neutralized according to the directions in the write-up or with a base using pH paper until a pH of about 7 is obtained. They can then be flushed down the drain with tap water. Waste from Part C should be collected in the proper waste container. All other waste can go down the sink with plenty of tap water.

EXPERIMENT 22: Le Châtelier's Principle

DATA SHEET 1

Part A: Solubility Equilibria

Tube	NaCl	H₂O	Total drops 12 M HCl	Observations
1	6 drops	0 drops	_____	_____
2	0 drops	6 drops	_____	_____

Part B: Acid–Base Equilibria (Indicators)

The available indicators may differ from those listed on the data page. Please alter the data below if necessary. Place 5 drops of H_2O in each well and test as follows:

Indicator (2 drops)	Acid Color (2 drops)	After Adding Base (6 drops)
1. Bromocresol Green	_____	_____
2. Thymol Blue	_____	_____
3. Bromchlorphenol Blue	_____	_____
4. Phenolphthalein	_____	_____

Indicator (2 drops)	Base Color (2 drops)	After Adding Acid (6 drops)
1. Bromocresol Green	_____	_____
2. Thymol Blue	_____	_____
3. Bromchlorphenol Blue	_____	_____
4. Phenolphthalein	_____	_____

EXPERIMENT 22: Le Châtelier's Principle

DATA SHEET 2

Part C: Complex Equilibria (INSTRUCTOR DEMO)

Six tubes, each containing 20 drops 0.3 M Co^{2+}, 12 drops H_2O, 45+ drops 12 M HCl

	Color	**Other Observations**
Original Solution	_____	_____
On Heating	_____	_____
On Cooling	_____	_____
On Adding 12 M HCl	_____	_____
On Adding $AgNO_3$	_____	_____
On Adding H_2O	_____	_____

EXPERIMENT 22: Le Châtelier's Principle

POSTLAB EXERCISE

1. Was the reaction in Part C an exothermic reaction or an endothermic reaction? Explain your reasoning.

Use the following information to answer the questions below:

When heat was added to the following reaction at equilibrium, 2NO was formed.

$$2NO(g) \rightleftharpoons N_2(g) + O_2(g)$$

2. In which direction will the reaction shift if $2NO(g)$ is added?

3. In which direction will the reaction shift with an increase of heat?

4. As written, is the reaction endothermic or exothermic?

5. What will happen to the value of K if heat is removed? Explain your reasoning.

EXPERIMENT 23: Acid–Base Properties of Hair Care Products

PRELAB EXERCISE

Terms:

Brønsted–Lowry acid –

Brønsted–Lowry base –

pH –

pH meter –

Buffer –

Safety Warnings:

EXPERIMENT 23: Acid–Base Properties of Hair Care Products

PRELAB EXERCISE

1. Why is it necessary to rinse the pH probe and blot it dry before transferring it from one solution to another?

23
ACID–BASE PROPERTIES OF HAIR CARE PRODUCTS

Objectives

1. To gain experience using a pH meter.

2. To learn about acid–base properties of hair care products.

3. To learn about buffers.

23.1 INTRODUCTION

A large part of the function of hair care products like shampoos and conditioners is determined by the acid–base properties of these products. A *Brønsted–Lowry acid* is a proton donor. A *Brønsted–Lowry base* is a proton acceptor. Since the definition of acid and base both use protons or H$^+$, a measurement of H$^+$ can tell us about acid–base properties. A way to measure the amount of H$^+$ in a sample is *pH*, which is defined as $-\log[\text{H}^+]$. A pH meter is a device used to measure pH. We will determine the acid–base properties of the hair care products using a pH meter.

These hair care products may contain several different compounds that have acid–base properties. These compounds will also have varying strengths as acids and bases. The stronger the acid or base the more completely it dissociates.

If a solution contains relatively large amounts of both an acid and its conjugate base, the solution is a buffer. Buffers resist change in pH when either a small amount of acid or base is added to them. By adding acid and base to the hair care products, we will be able to determine if a buffer is present. (Assume that a buffer is present when the pH change is less than 1 for both the acid and base additions.) Deionized water will be tested to show how an unbuffered solution will respond to an acid and a base.

23.2 EXPERIMENTAL (WORK IN PAIRS)

To calibrate the pH meter:

The probe should remain in the flexible electrode support arm during use.

1. Remove the probe tip from the storage solution.

2. Rinse the probe tip with deionized water into a waste beaker. Blot dry with a lint-free tissue.

3. Place the probe tip into a buffer solution (usually pH 7 is used); gently swirl and press the standardization key. While standardizing, the meter's readout will have "standardizing" displayed. After the meter has stabilized the readout will say "measuring." When a stable measurement is reached an "S" icon will be displayed.

4. The probe may be removed from the buffer and rinsed, blotted, and placed in the storage solution to await student use. Remember to **gently** swirl.

To use the pH meter:

1. Make sure the meter is plugged in and calibrated. Note: Your instructor will do the calibration step.

2. After the meter has been calibrated, remove the probe tip from the storage solution.

3. Rinse the probe tip with deionized water into a waste beaker. Blot dry with a lint-free tissue.

4. To measure pH, place the probe tip into a solution and gently swirl. The display will give an **S** icon when the reading has stabilized.

5. Remove the probe tip, rinse into the waste beaker with deionized water, blot with a lint-free tissue, and place in the next solution or the storage solution.

Preparation of Hair Care Products

Clean and dry twenty 50 mL beakers.

Place a 10 mL sample of hair care product into one of the beakers. Label the beaker for the product; also include an "A" on the label. Make sure that the label does not cover the volume markings on the beaker. Gently pour enough deionized water into the beaker to give a final volume of 40 mL. Use a clean glass stirring rod to mix the solution. Once mixed thoroughly, pour half of the content into another clean beaker and label it for the product and include a "B" on the label. Make sure that the label does not cover the volume markings on the beaker. Repeat this process until you have nine samples from different hair care products and one sample of deionized water for reference. Record the product names in the data table.

pH Measurements of Hair Care Products

Measure and record the pH for each solution.

pH Changes Caused by Adding Strong Acid or Strong Base to Hair Care Products

To the first hair care product beaker labeled "A" add 3 drops of 0.1 M HCl. Mix, measure, and record the pH. Add 3 more drops of 0.1 M HCl. Mix, measure, and record the pH.

To the first hair care product beaker labeled "B" add 3 drops of 0.1 M NaOH. Mix, measure, and record the pH. Add 3 more drops of 0.1 M NaOH. Mix, measure, and record the pH.

Repeat this process for each of the other samples.

Determine the pH change that each product experiences upon addition of both the acid and base.

Safety Notes

HCl and NaOH should be treated with care.

23.3 WASTE DISPOSAL

Waste for this lab may be dissolved in tap water and flushed down the sink.

EXPERIMENT 23: Acid–Base Properties of Hair Care Products

DATA SHEET 1

Beaker Contents:

DI H$_2$O

Divide into two equal volumes

	1A	1B	2A	2B	3A	3B	4A	4B	5A	5B
Check the pH$_{initial}$	___	___	___	___	___	___	___	___	___	___
Add 3 drops	0.1 M HCl	0.1 M NaOH	0.1 M HCl	0.1 M NaOH	0.1 M HCl	0.1 M NaOH	0.1 M HCl	0.1 M NaOH	0.1 M HCl	0.1 M NaOH
Check the pH	___	___	___	___	___	___	___	___	___	___
Add 3 drops	0.1 M HCl	0.1 M NaOH	0.1 M HCl	0.1 M NaOH	0.1 M HCl	0.1 M NaOH	0.1 M HCl	0.1 M NaOH	0.1 M HCl	0.1 M NaOH
Check the pH$_{final}$	___	___	___	___	___	___	___	___	___	___
ΔpH $(pH_{final} - pH_{initial})$	___	___	___	___	___	___	___	___	___	___

EXPERIMENT 23: Acid–Base Properties of Hair Care Products

DATA SHEET 2

Beaker Contents:

Divide into two equal volumes

	6A 6B	7A 7B	8A 8B	9A 9B	10A 10B

Check the pH$_{initial}$ ___ ___   ___ ___   ___ ___   ___ ___   ___ ___

Add 3 drops	0.1 M HCl	0.1 M NaOH	0.1 M HCl	0.1 M NaOH	0.1 M HCl	0.1 M NaOH	0.1 M HCl	0.1 M NaOH	0.1 M HCl	0.1 M NaOH

Check the pH   ___ ___   ___ ___   ___ ___   ___ ___   ___ ___

Add 3 drops	0.1 M HCl	0.1 M NaOH	0.1 M HCl	0.1 M NaOH	0.1 M HCl	0.1 M NaOH	0.1 M HCl	0.1 M NaOH	0.1 M HCl	0.1 M NaOH

Check the pH$_{final}$ ___ ___   ___ ___   ___ ___   ___ ___   ___ ___

ΔpH
(pH$_{final}$ *- pH*$_{initial}$*)*   ___ ___   ___ ___   ___ ___   ___ ___   ___ ___

EXPERIMENT 23: Acid–Base Properties of Hair Care Products

POSTLAB EXERCISE

1. Based upon your results, which of the hair care products contain buffers?

2. Referencing the labels for the products listed in question 1, which compounds would make up the buffer?

EXPERIMENT 24: Brønsted–Lowry Acids and Bases

PRELAB EXERCISE

Terms:

      Brønsted–Lowry acid –

      Brønsted–Lowry base –

      Conjugate acid –

      Conjugate base –

      pH –

      pOH –

Safety Warnings:

EXPERIMENT 24: Brønsted–Lowry Acids and Bases

PRELAB EXERCISE

1. Write a net ionic equation that describes a slightly basic solution resulting from dissolving $NaC_2H_3O_2$ in water.

2. Write the conjugate acid for each of the following:

    a. NH_3_____

    b. $C_2H_3O_2^-$_____

    c. OH^-_____

3. A solution has a hydronium concentration of 7.3×10^{-5} M. Calculate:

    a. pH

    b. pOH

    c. $[OH^-]$

4. Is the solution in question 3 acidic, basic, or neutral?

BRØNSTED–LOWRY ACIDS AND BASES

Objectives

1. To gain experience using pH meters.

2. To study the effects of adding acids, bases, and neutral solutions to a Brønsted–Lowry acid.

3. To study the effects of adding acids, bases, and neutral solutions to a Brønsted–Lowry base.

4. To gain experience writing net ionic equations.

24.1 INTRODUCTION

A very useful concept of protonic acids was introduced by Brønsted and Lowry in which acids are defined as proton donors, bases as proton acceptors, and an acid–base reaction as the transfer of a proton, H^+, from an acid to a base. The generalized Brønsted–Lowry (B–L) acid–base reaction can be written as:

$$HA + B \rightleftharpoons A^- + HB^+$$

where HA and HB^+ are acids and A^- and B are, respectively, their conjugate bases. The equilibrium may be studied either by mixing solutions of HA and B, or solutions of HB and A. Thus, all B–L reactions may be viewed as a competition between two bases (here, B and A^-) for the proton. The resulting reaction depends upon the relative strengths of the two competing bases—that is, if A is a better proton-acceptor (stronger base) than B, then the reverse reaction (right to left) is favored and vice versa.

It follows that in the B–L concept:

(a) The products of an acid–base reaction are a new acid and a new base.

(b) The stronger an acid, the weaker its conjugate base; the reverse is also true, the stronger the base, the weaker its conjugate acid.

(c) The preferred direction of net chemical change is toward formation of the weaker acid and weaker base.

A substance commonly used either as B–L acid or B–L base is water, and in practice it is used as a reference acid or base for added solutes (the three-line equal mark indicates a definition):

$$HA(aq) + H_2O(l) \rightleftharpoons H_3O^+ + A^- \qquad K \equiv K_a \text{ for HA}$$

$$B(aq) + H_2O(l) \rightleftharpoons HB^+ + OH^- \qquad K = K_b \text{ for B}$$

Substances added to water are acids if they increase the concentration of the hydronium ion, H_3O^+. Bases increase the concentration of the hydroxide ion, OH^-. This categorization can be done by determining the pH of water, adding the solute, and then determining the pH of the solution. Water is usually slightly acidic due to traces of dissolved CO_2. (Why should this be so?)

The following information applies to pH:

- pH = $-\log [H_3O^+]$, or alternatively, $[H_3O^+] = 10^{-pH}$.

- Pure water (25 °C), an exactly neutral solution, has a pH = 7, with $[H_3O^+] = [OH^-]$.

- An acidic solution has a pH < 7, with $[H_3O^+] > [OH^-]$.

- An alkaline (basic) solution has a pH > 7, with $[H_3O^+] < [OH^-]$.

- Solutes that cause a pH *decrease* are acids.

- Solutes that cause a pH *increase* are bases.

- A pH change of x units corresponds to a 10 change in hydronium ion concentration (a slight pH change can mean a large change in $[H_3O^+]$).

- The pH of a solution can be measured electronically using a pH meter or a portable device (used here) called a digital pH probe.

This experiment involves the use of pH measurements to distinguish aqueous solutes as acids or bases. Or, if either an acid or a base is added in slight excess of the stoichiometric proportions then sudden abrupt pH changes will occur. This will also be detected with use of the digital pH probe.

24.2 EXPERIMENTAL (WORK IN PAIRS)

Part A:

Clean and dry eleven 50 mL beakers.

In six separate 50 mL beakers, obtain about 20 mL of each of the following 0.6 M solutions: NaOH (strong base, solution of OH^-), HCl (strong acid, solution of H_3O^+), NH_3, $NaC_2H_3O_2$, NaCl, and $NaHSO_4$.

Set aside five clean 50 mL beakers; five clean, small-diameter stirring rods; a pH probe; and several droppers. (If only one stirring rod is used it must be rinsed and dried before use with a different solution.) Since the solutions are equimolar, equal volumes will contain equal numbers of moles. Place 15 mL of deionized water into each beaker, and label the beakers 1 through 5.

Please note: After each measurement and standardization, the pH probe must be rinsed with deionized water and blotted dry with a lint-free tissue. Test the pH meter by placing it into the buffer solution provided to ensure that it again reads ± 0.2 pH units of the accepted value. Please handle the pH probe very carefully!!

Test 1

In beaker 1, place 6 mL of 0.6 M Reagent A, stir with a clean stirring rod, and determine and record the pH. Now, add 2 mL of 0.6 M Reagent B, stir and determine the solution pH (record). Repeat with 4 additional mL of 0.6 M Reagent B and then again with 4 more mL, determining and recording the pH after each addition. Carry out tests as in the preceding paragraph with the combinations of Reagents A and B below:

Test 1: Reagent A = 0.6 M NaOH   Reagent B = 0.6 M $NaHSO_4$

Test 2: Reagent A = 0.6 M HCl   Reagent B = 0.6 M NH_3

Test 3: Reagent A = 0.6 M HCl   Reagent B = 0.6 M $NaC_2H_3O_2$

Test 4: Reagent A = 0.6 M HCl   Reagent B = 0.6 M NaCl

Test 5: Reagent A = 0.6 M NaOH   Reagent B = 0.6 M NaCl

Upon completion of these tests, clean and rinse the beakers, stirring rods, and droppers. Remove the labels from the beakers.

Part B: pH of Various Solutes (Instructor Demonstration)

Using the same pH meter to reduce error, determine the pH for each of the following solutions: 0.6 M $NaC_2H_3O_2$, 0.6 M NaCl, and 0.6 M $NaHSO_4$. Indicate whether each of the aqueous solutes is an acidic, basic, or neutral solute.

Add a spatula tip-full of solid calcium hydride, CaH_2, to a 100 mL beaker containing 20 mL of deionized water and then test the pH of the resulting solution. Which solution from the previous paragraph ($NaC_2H_3O_2$, NaCl, $NaHSO_4$) should give the most pronounced acid–base reaction with this solution? (This would be the solution with the pH most different from the CaH_2 solution.) Test your prediction by adding 1 mL of the appropriate 0.6 M solution and testing the pH. Add several additional milliliters and check the pH again. Note any other observations that would seem appropriate.

Safety Notes

HCl and NaOH should be treated with care. Calcium hydride releases flammable gas when it comes into contact with water. Use ventilation when carrying out the procedure in Part B, and to minimize the reaction, use only the small amount indicated in the instructions.

24.3 Waste Disposal

All solutions may be flushed down the drain with plenty of tap water.

EXPERIMENT 24: Brønsted–Lowry Acids and Bases

DATA SHEET 1

Part A

Test	1	2	3	4	5
15 mL	H_2O	H_2O	H_2O	H_2O	H_2O
6 mL	NaOH	HCl	HCl	HCl	NaOH
Initial pH	_____	_____	_____	_____	_____
Add 2 mL	B. ↓ $NaHSO_4$	B. ↓ NH_3	B. ↓ $NaC_2H_3O_2$	B. ↓ NaCl	B. ↓ NaCl
Check pH	_____	_____	_____	_____	_____
Add 4 mL	B. ↓ $NaHSO_4$	B. ↓ NH_3	B. ↓ $NaC_2H_3O_2$	B. ↓ NaCl	B. ↓ NaCl
Check pH	_____	_____	_____	_____	_____
Add 4 mL	B. ↓ $NaHSO_4$	B. ↓ NH_3	B. ↓ $NaC_2H_3O_2$	B. ↓ NaCl	B. ↓ NaCl
Check pH	_____	_____	_____	_____	_____

EXPERIMENT 24: Brønsted–Lowry Acids and Bases

DATA SHEET 2

Part B: pH of Various Solutes

A. $NaC_2H_3O_2$ _____

B. NaCl   _____

C. $NaHSO_4$ _____

D. CaH_2/H_2O _____

Observations on dissolving CaH_2 _____

Which solute (A, B, or C) would have the most pronounced effect on the pH of solution D? _____

pH after adding 1 mL = _____

pH after adding 5 more mL = _____

pH after adding 10 more mL = _____

Other observations _____

EXPERIMENT 24: Brønsted–Lowry Acids and Bases

POSTLAB EXERCISE

1. Write the net ionic equation for the CaH_2/H_2O reaction in Part B.

2. Write the net ionic equation for the products of the reaction in question 1 reacting with the solution you chose in Part B ($NaC_2H_3O_2$, NaCl, $NaHSO_4$).

3. A few mL of _____ M HCl are added to 20 mL of _____ M $NaC_2H_3O_2$:

   a. Will the initial pH of the _____ M $NaC_2H_3O_2$ be greater than, less than, or equal to 7?

      A. Greater than 7   B. Less than 7   C. Equal to 7

   b. As HCl is added to the _____ M $NaC_2H_3O_2$, will the pH of the solution increase, decrease, or remain unchanged?

      A. Increase   B. Decrease   C. Remain unchanged

   c. Write the net ionic equation for the reaction between the HCl solution and the $NaC_2H_3O_2$ solution.

4. A solution has a hydroxide concentration of _____ M. Calculate the following, showing your work.

   a. pH

   b. pOH

   c. $[H^+]$

EXPERIMENT 25: Buffers

PRELAB EXERCISE

Terms:

Buffer –

K_b –

K_a –

Safety Warnings:

EXPERIMENT 25: Buffers

PRELAB EXERCISE

1. Weak acid A has a K_a of 1.1×10^{-4}. What is the pH of a solution that is 0.75 M in the weak acid A and 0.25 M in its conjugate base?

2. Weak acid C has a K_a of 1.8×10^{-4}. What is the pH of a solution that is 1.0 M in both the weak acid C and its conjugate base?

3. Which is the "stronger" acid, A or C? Explain your answer.

BUFFERS

Objectives

1. To gain experience using a pH meter.

2. To learn about buffers.

3. To be able to calculate the pH of a buffer solution and predict how it will react to addition of an acid or base.

25.1 INTRODUCTION

Even in quite dilute solutions, acetic acid is very slightly ionized (it would approach 99% ionization only as the concentration approaches 0.0M):

$$HC_2H_3O_2(aq) + H_2O(l) \rightleftharpoons H_3O^+(aq) + C_2H_3O_2^-(aq) \quad K_a = 1.8 \times 10^{-5}$$

On the other hand, the salt, sodium acetate trihydrate ($NaC_2H_3O_2 \cdot 3H_2O$), is essentially 100% dissociated into the constituent hydrated ions in a dilute aqueous solution. Sodium acetate solutions are basic because the acetate ion (the conjugate base of acetic acid) behaves as a proton acceptor with respect to water:

$$C_2H_3O_2^-(aq) + H_2O(l) \rightleftharpoons OH^-(aq) + HC_2H_3O_2(aq) \quad K_b = 5.6 \times 10^{-10}$$

Consequently, in solutions of acetic acid alone, the molar concentration of the $HC_2H_3O_2$ is much larger than the $C_2H_3O_2^-$ concentration, but in solutions of sodium acetate alone, the reverse is true. Neither solute alone can provide "comparable" concentrations in solutions of both the weak acid and its conjugate base; thus neither an acetic acid solution nor a sodium acetate solution is a buffer solution. A buffer solution, by definition, *must contain moderate concentrations of both species!*

The following is true of buffers:

- The pH of a solution of a weak acid (if not extremely weak) is governed by the concentration of the acid and K_a.

- The pH of a solution of a weak base is determined by the concentration of the weak base and K_b.

- In a solution containing both a weak acid and a strong acid, both acids play a role in determining the pH of the solution.

- If in the solution containing both a weak acid and a strong acid, the concentration of the strong acid is relatively large, it will <u>inhibit</u> the dissociation of the weak acid (the common-ion effect). The pH of this solution would then be calculated as if the weak acid were not present!

- In a solution containing both a strong base and a weak base, the strong base concentration would be used to calculate pH.

Buffer solutions contain both a weak acid and its conjugate weak base in appreciable concentrations. Within limits, these solutions tend to resist changes in pH upon addition of either H_3O^+ or OH^- (because these species are largely consumed by the acidic and basic components of the buffer mixture). In buffer systems like $NaC_2H_3O_2 \cdot HC_2H_3O_2$ mixtures, the principal source of the acetic acid molecule is from the acid; the principal source of the acetate ion is from the salt. Therefore, the $[H_3O^+]$ is determined by the salt/acid (or equivalently, the base/acid) mole ratio. For the acetic acid–acetate system,

$$[H_3O^+] = K_a \frac{[HC_2H_3O_2]}{[C_2H_3O_2^-]} = K_a \frac{n_{HC_2H_3O_2}}{n_{C_2H_3O_2^-}} = K_a \frac{n_{acid}}{n_{conjugate\ base}}$$

where n stands for moles (or millimoles). Taking negative logarithms of both sides and using the symbol, p, to indicate "$-\log_{10}$ of"

$$pH = pK_a + \log_{10} \frac{n_{conjugate\ base}}{n_{acid}}$$

For acetic acid at 25 °C, $pK_a = 4.74$. Show how this number is related to the K_a for acetic acid given above.

This experiment deals with the pH of a solution of acetic acid alone and with a solution of sodium acetate alone (Part A); of "direct" buffer solutions prepared by mixing solutions of acetic acid with solutions of sodium acetate (Part B); of "indirect" buffers prepared by partially "neutralizing" an acetic acid solution with NaOH (Part C), and with the comparison of the pH changes observed when strong acid (HCl) and strong base (NaOH) solutions are added to equal volumes of buffered and unbuffered solutions (Part D).

25.2 EXPERIMENTAL (WORK IN PAIRS)

Note: You will use the same pH probes as in the previous experiment. Please remember that the probe must be thoroughly rinsed and recalibrated using the pH = 7 buffer periodically.

Part A: pH Measurements on Solutions of Acetic Acid and Sodium Acetate

Acquire 100 mL of 0.60 M acetic acid from your instructor. Also, prepare 100 mL of 0.60 M sodium acetate trihydrate, $NaC_2H_3O_2 \cdot 3H_2O$, by dissolving the solid salt provided and likewise record all pertinent data. Transfer 10–20 mL of these solutions to separate, clean 50 mL beakers; measure and record the pH of each solution—these data will be used to compare the observed pH with those calculated using the K_a and K_b values provided.

Part B: pH Measurements on Buffer Mixtures of Acetic Acid and Sodium Acetate

Using the 0.60 M solutions prepared in Part A, prepare in clean 50 mL beakers the following ("direct") buffer solutions:

Beaker	mL 0.60 M Acetic Acid	mL 0.60 M NaC$_2$H$_3$O$_2$
1	16	4
2	12	8
3	10	10
4	6	14
5	2	18

Measure the pH and record the results for each solution. These will be used to compare with those calculated using pK_a.

Part C: pH Measurements on Solutions Prepared by Reaction of Acetic Acid with Sodium Hydroxide

Place 20 mL of the 0.60 M acetic acid in a clean 50 mL beaker. Measure and record the pH after adding the following volumes of 1.0 M NaOH (Be sure to mix the solution after each addition.):

Addition	Total mL 1.0 M NaOH added
1	2
2	5
3	10
4	20

Do all of these additions result in buffer solutions prepared (indirectly) by partial "neutralization" of the acid? The measured results on these solutions will be used to compare with the calculated values.

This solution must be neutralized before disposal. See instructions in the Waste Disposal section at the end of the write-up.

Part D: pH Changes Caused by Adding Strong Acid and Strong Base to Buffered and Unbuffered Solutions

In clean 50 mL beakers, prepare the following solutions and record the initial pH of each:

Beaker	Contents
1	10 mL 0.60 M $HC_2H_3O_2$ + 10 mL H_2O
2	10 mL 0.60 M $NaC_2H_3O_2$ + 10 mL H_2O
3	20 mL 0.30 M NaCl
4	10 mL 0.60 M $HC_2H_3O_2$ + 3 mL 1.0 M NaOH + 7 mL H_2O
5	20 mL H_2O

Divide each of these solutions into two equal volumes. Label these as 1A (for acid addition) and 1B (for base addition), 2A and 2B, and so forth.

Caution: The 6.0 M HCl and NaOH can cause burns. Rinse your hands with water and wipe up spills if they occur!

Now add 3 drops (only!) of 6.0 M HCl to each beaker labeled A, stir, and remeasure (record) the pH.

To each beaker labeled B, add 3 drops (only!) of 6.0 M NaOH, stir, and remeasure (record) the pH. To neutralize the solutions, combine the contents of the 10 beakers in a 100 mL beaker.

Explain these results (pH changes) in terms of the buffered and unbuffered solutions in beakers 1 to 5.

Safety Notes

The 6 M HCl and 6 M NaOH are very corrosive. Avoid contact with eyes and skin. See *Caution* given above. The 1 M NaOH is somewhat corrosive. Handle all chemicals with care.

25.3 WASTE DISPOSAL

The solutions can be flushed down the drain with a copious amount of tap water.

EXPERIMENT 25: Buffers

DATA SHEET 1

Part A

100 mL 0.6 M
acetic acid
pH = _____

100 mL 0.6 M
sodium acetate
pH = _____

Part B

Beaker	#1	#2	#3	#4	#5
0.6 M HOAc	16 mL	12 mL	10 mL	6 mL	2 mL
0.6 M NaOAc	4 mL	8 mL	10 mL	14 mL	18 mL
pH =	_____	_____	_____	_____	_____

Part C

Addition	#1	#2	#3	#4
0.6 M HOAc	20 mL	20 mL	20 mL	20 mL
1.0 M NaOH	2 mL	+ 3 mL = 5 mL	+ 5 mL = 10 mL	+ 10 mL = 20 mL
pH =	_____	_____	_____	_____

Note: The # is used for clarity in the table above.

EXPERIMENT 25: Buffers

DATA SHEET 2

Part D

Beaker	1	2	3	4	5
	10 mL 0.60 M $HC_2H_3O_2$	10 mL 0.60 M $NaC_2H_3O_2$	20 mL 0.30 M NaCl	10 mL 0.60 M $HC_2H_3O_2$ + 3 mL 1.0 M NaOH	20 mL H_2O
	+ 10 mL H_2O	+ 10 mL H_2O		+ 7 mL H_2O	

Initial pH ___ ___ ___ ___ ___

Divide into two equal volumes

	1A	1B	2A	2B	3A	3B	4A	4B	5A	5B
Add 3 drops ONLY	6.0 M HCl	6.0 M NaOH	6.0 M HCl	6.0 M NaOH	6.0 M HCl	6.0 M NaOH	6.0 M HCl	6.0 M NaOH	6.0 M HCl	6.0 M NaOH

Check the pH ___ ___ ___ ___ ___ ___ ___ ___ ___ ___

EXPERIMENT 25: Buffers

DATA SHEET 3

Part A: Initial Solutions

_____ g $NaC_2H_3O_2$, _____ mL solution

	pH measured	**pH calculated**
0.60 M Acetic Acid	_____	_____
0.60 M Sodium Acetate	_____	_____

Part B: Buffers Direct Method

Beaker	**mL $HC_2H_3O_2$**	**mL $NaC_2H_3O_2$**	**pH measured**	**pH calc.**
1	16	4	_____	_____
2	12	8	_____	_____
3	10	10	_____	_____
4	6	14	_____	_____
5	2	18	_____	_____

Part C: Buffers Indirect Method

Addition of 1.0 M NaOH to 20 mL of 0.60 M $HC_2H_3O_2$

Addition	**Total mL 1.0 M NaOH added**	**pH measured**	**pH calc.**
1	2	_____	_____
2	5	_____	_____
3	10	_____	_____
4	20	_____	_____

EXPERIMENT 25: Buffers

DATA SHEET 4

Part D: pH Changes

I. *Addition of 3 drops of 6 M HCl*

Beaker	Contents	pH initial	pH final	ΔpH
1	_____	_____	_____	_____
2	_____	_____	_____	_____
3	_____	_____	_____	_____
4	_____	_____	_____	_____
5	_____	_____	_____	_____

II. *Addition of 3 drops of 6 M NaOH*

Beaker	Contents	pH initial	pH final	ΔpH
1	_____	_____	_____	_____
2	_____	_____	_____	_____
3	_____	_____	_____	_____
4	_____	_____	_____	_____
5	_____	_____	_____	_____

EXPERIMENT 25: Buffers

POSTLAB EXERCISE

1. Discuss the difference between a weak acid (A) and a strong acid (SA).

2. If a solution is 1 M in the weak acid A and another solution is 1 M in the strong acid SA, which solution will have the lowest pH? Explain your reasoning.

3. List the solutions in Part D that are buffers.

EXPERIMENT 26: Gas Chromatography and Identification of an Unknown

PRELAB EXERCISE

Terms:

Chromatography –

Chromatogram –

Moving phase –

Stationary phase –

Gas chromatography –

Retention time –

Safety Warnings:

EXPERIMENT 26: Gas Chromatography and Identification of an Unknown

PRELAB EXERCISE

1. From a mixture, on what basis can the substances be separated using the technique of chromatography?

26

GAS CHROMATOGRAPHY AND IDENTIFICATION OF AN UNKNOWN

Objectives

1. To gain experience using gas chromatography.

2. To use gas chromatography to identify an unknown.

26.1 INTRODUCTION

Chromatography is a technique used to separate and identify small quantities of mixtures into their component parts. This technique was so named due to the highly colored (Greek, *chroma*, colored) components separated from chloroplast extracts when this technique was first used in 1906 by M. Tswett.

All forms of chromatography employ two phases; a *moving phase* and a *stationary phase*. The moving phase is a solvent, or mixture of solvents, that flows over the other material, the stationary phase. In chromatography, mixtures can be separated because each substance in the mixture will have different affinities for the stationary and moving phases. When the correct stationary and moving phases are selected the mixture separates because:

- *each substance in the mixture is adsorbed on the stationary phase with a different degree of tenacity.*

- *each substance in the mixture has a different affinity for the moving phase.*

A compound with a high affinity for the moving phase and a low affinity for the stationary phase will move through the *stationary phase* rapidly. A compound with a high affinity for the stationary phase and a low affinity for the moving phase will move through the *stationary phase* slowly.

Gas chromatography utilizes a gas as the moving phase. The carrier gas for the gas chromatograph (GC) in lab this week is helium. There is a column containing a stationary phase that the carrier gas passes through. A mixture is injected with a syringe and is carried to the stationary phase by the carrier gas. The closer a component in a mixture is to the stationary phase, the more time it will spend on the stationary phase. This means it will take longer for the substance to exit the column and be detected.

This time required for the mobile phase to move a solute from its injection point, through the stationary phase, and to the detector is the *retention time*. Retention time can be used to identify components of a mixture by comparing a solute's retention time with that of a known compound. If the retention times are different, the compounds are different. If the retention times are the same, there is evidence that the compounds are the same.

When trying to positively identify a component of a mixture, more information is needed. This can be obtained in several ways. One, several different types of stationary phase/mobile phase combinations can be used. Usually three different combinations offer sufficiently different conditions to be able to determine the identity of the unknown. It is extremely unlikely that two different chemicals will respond the same under the three different conditions. Two, extra information can be obtained. For example, the detector can be a mass spectrometer or an infrared spectrometer that can provide extra information about the identity of the component. To keep this experiment as simple as possible, we will not be using these extra techniques.

26.2 EXPERIMENTAL (WORK IN GROUPS)

The following should be located at the back of the laboratory room.

– syringe

– 3 GCs labeled GC#1, GC#2, and GC#3

– 1 computer with data acquisition software

– 1 printer

– 1 sample vial with 1-propanol

– 1 sample vial with 2-propanol

– 1 sample vial with iso-butanol (also known as 2-methyl-1-propanol)

– 1 sample vial with a mixture of 1-propanol, 2-propanol, iso-butanol

– 1 sample vial with your unknown (Note: Your unknown is some combination of one or more of 1-propanol, 2-propanol, and iso-butanol.)

The class will be broken into three groups. Each group will be assigned to a specific GC. When using the computer, it will reference your GC by its number. For example, GC#1 will be referenced as GC 1 or Channel 1.

Your lab instructor will demonstrate how to use the computer to acquire data from the GC, how to make an injection into the GC, and how to print out the results of your trials. Each trial requires 10 minutes to complete. You will need to make efficient use of your time. When finished with each trial, print out enough chromatograms for each member of your group to have one for their laboratory report. Make sure each chromatogram is labeled with exactly what is injected.

Trial 1:   Inject 2 μL of 1-propanol into the GC and start data collection for that GC at the same time. Print out the chromatograms.

Trial 2:   Inject 2 μL of 2-propanol into the GC and start data collection for that GC at the same time. Print out the chromatograms.

Trial 3:   Inject 2 μL of iso-butanol into the GC and start data collection for that GC at the same time. Print out the chromatograms.

Trial 4:   Inject 4 μL of the mixture of 1-propanol, 2-propanol, iso-butanol into the GC and start data collection for that GC at the same time. Print out the chromatograms. You should be able to identify each of the components of the mixture by its retention time.

Trial 5: Inject 4 µL of your unknown into the GC and start data collection for that GC at the same time. Print out the chromatograms. Identify the unknown based upon the retention time.

You will have to turn in a copy of the chromatograms.

Safety Notes

Be careful using the needles.

26.3 Waste Disposal

Return all chemicals to your lab instructor.

EXPERIMENT 26: Gas Chromatography and Identification of an Unknown

DATA SHEET

Unknown ID: _____

Boiling Point Data: (use reference material like the *Merck Index* or the *CRC Handbook*)

   – 1-propanol   _____°C

   – 2-propanol   _____°C

   – iso-butanol   _____°C

   Reference used:

Retention Time: (include units)

   – 1-propanol   _____

   – 2-propanol   _____

   – iso-butanol   _____

   – Mixture:

         • 1-propanol   _____

         • 2-propanol   _____

         • iso-butanol _____

   – Unknown   _____

                    _____

                    _____

EXPERIMENT 26: Gas Chromatography and Identification of an Unknown

POSTLAB EXERCISE

1. What is your unknown identity number?

2. What is the identity of the component(s) in your unknown? (5 points)

3. Explain how you were able to determine the identity of the component(s) in your unknown. (5 points)

ELECTRONIC BALANCE

Use of the Electronic Balance

Note: To reduce experimental error, use the same balance when taking multiple measurements.

1. Make sure that the balance is clean before use to prevent your sample from becoming contaminated.

2. Place weigh paper on the balance pan to protect its surface.

3. Press the "O/T" button to "turn on" the balance and "tare" its reading to 0.00 g.

4. Place your sample on the weigh paper.

5. Allow reading to stabilize and record.

6. Remove sample.

7. To prevent contamination, place weigh paper in appropriate waste container.

8. Make sure balance is clean.

9. If you are the last user of the day, press the "Mode/Off" switch until the balance turns off.

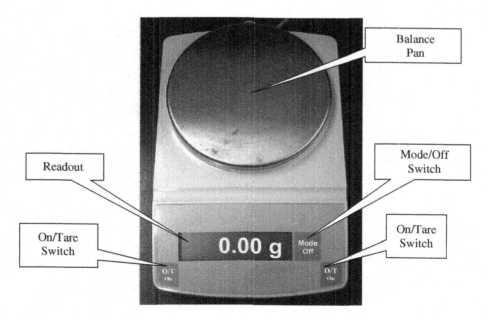

B

BUNSEN BURNER

To Operate a Bunsen Burner

1. Inspect the burner tubing for cracks and connect one end to the gas supply valve and the other end to the gas inlet of the burner.

2. Open the needle valve at least a half turn.

3. When you are ready to light the flame (with a striker or a match), turn on the gas supply and light at the end of the barrel.

4. The fuel going to the flame can be adjusted with the needle valve. Take care not to completely unscrew the needle valve.

5. The air going to the flame can be adjusted by turning the barrel. Take care not to completely unscrew the barrel.

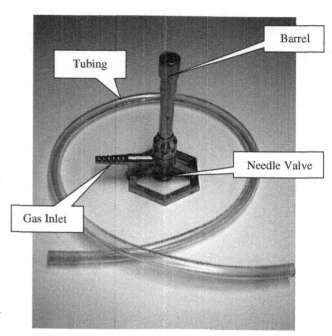

6. As you increase the air supply, more gas will be required to give a stable flame.

7. The most useful flame burns relatively quietly and has two distinct zones; the inner one is cone-shaped and light blue. The tip of the inner cone is the hottest point of the flame.

8. The height and nature of the flame should be adjusted to suit the nature of the job.

9. When you no longer need the flame, first turn off the gas at the gas supply, and then close the needle valve.

10. *BUNSEN BURNER SAFETY WARNING: SHOULD THE FLAME NOT LIGHT OR GO OUT, OR SOMETHING UNEXPECTED HAPPENS, TURN OFF THE GAS AT THE GAS SUPPLY VALVE ON THE BENCHTOP! MAKE SURE THAT YOU TURN OFF THE GAS AT THE BENCHTOP GAS SUPPLY VALVE WHEN YOU NO LONGER NEED THE FLAME.*

C

VOLUMETRIC PIPET

Use of the Volumetric Pipet

1. Make sure that the pipet is clean before use. If not, use soap and water to clean it. Make sure to rinse thoroughly with water to remove all soap.

2. Rinse the pipet with deionized water from a wash bottle. Allow all the water to drain out the tip into a waste beaker. There will probably be a drop of water left in the tip of the pipet. Use the pipet bulb to gently force this water into the waste beaker.

3. Rinse the pipet with the solution that will be transferred. To do this, place a small amount of the solution in a clean beaker. Place the tip of the pipet below the liquid level in the beaker. Using a bulb or pipettor, draw liquid into the pipet until it begins to fill the thick region of the pipet. Do not allow the liquid to be drawn up into the bulb or pipettor. If that happens shake the bulb or pipettor as dry as possible. If you are working with any liquid other than water you will need to rinse the liquid from the bulb before shaking the bulb dry. Remove the bulb or pipettor and place your finger over the end of the pipet to stop the liquid flow. Lift the pipet tip out of the liquid and turn the pipet sideways. Rotate the pipet to allow the liquid to coat the inside of the pipet. Allow the liquid that will to drain out into a waste beaker. There will probably be a drop of liquid left on the tip of the pipet. Touch just the drop to the waste beaker and it will drain from the pipet. Repeat this rinsing process for a total of three times.

4. You are now ready to fill the pipet. Place the solution to be transferred into a clean beaker. Place the tip of the pipet below the liquid level in the beaker.

    a. If using a bulb: squeeze the bulb and place it on the end of the pipet. Release pressure on the bulb to allow the liquid to flow into the pipet. Allow the liquid to rise above the calibration line. Quickly remove the bulb and place your finger over the end of the pipet to stop the liquid from draining. Remove the tip of the pipet from the liquid. While holding the pipet as vertically as possible and with your eyes level with the calibration line, slowly reduce pressure on the tip of the pipet to allow the liquid to drain so that the bottom of the meniscus is just touching the calibration line. Use your finger to stop the liquid flow at this point. Be patient; it may take multiple attempts.

    b. If using a pipettor: place the open end of the pipettor on the end of the pipet. Use the roller to allow liquid to flow into the pipet. Adjust the roller until the bottom of the meniscus is just touching the calibration line.

5. Remove the pipet from the beaker. Gently touch the tip of the pipet against the liquid in the beaker to drain off any excess liquid at the tip.

6. Move the tip of the pipet into the container you want to transfer to. Remove your finger or the pipettor to allow the liquid to drain.

7. Touch the tip of the pipet to the liquid you have just transferred to remove any accounted for liquid.

8. Most volumetric pipets are designed to drain the specified volume and they are labeled "TD" meaning "to deliver." If there is any liquid left in the pipet at this point it is excess and can be placed in the appropriate waste container.

9. Clean the pipet after use and rinse with deionized water.

D

BURETTE

Setup of the Burette

1. Make sure that the burette is clean before use. If not, use soap and water to clean it. Make sure to rinse thoroughly with water to remove all soap.

2. Rinse the burette with deionized water from a wash bottle. To minimize the use of wash water, tilt the burette and rotate the barrel while rinsing. This uses a minimum of wash water to coat the inside of the burette. Allow the water to drain into a waste beaker.

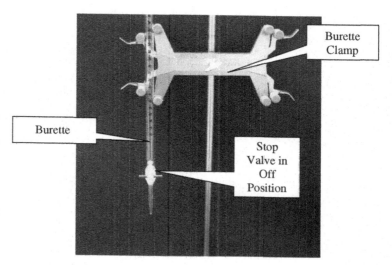

3. Rinse the burette with the solution that will be used. To do this, place a small amount of the solution in a clean beaker. Slowly pour 3 to 4 mL of solution into the burette while tilting and rotating the barrel to fully coat the inside. This uses a minimum of solution to coat the inside of the burette. Allow the solution to drain into a waste beaker. Repeat this rinsing process for a total of three rinses.

4. Place the burette securely in the burette clamp as shown in the figure. It should be vertical in the clamp. Place a clean funnel over the opening at the top of the burette. Place the waste beaker under the burette. Pour some of the solution to be used into a clean beaker.

5. You are now ready to fill the burette. Lower the burette in the clamp until the funnel is no higher than eye level. Pour from the beaker into the funnel to partially fill the burette.

6. Drain some of the solution to fill the burette tip and stop valve with solution. Note: Make sure that there are no air bubbles in the tip and stop valve!

7. Fill the burette with the solution that will be used from the beaker. You can drain it to between 0.00 and 1.00 mL.

8. All burette readings should be taken to two decimal places.

9. When reading the burette, never allow the solution in the burette to drain below the 50 mL mark on the burette.

10. Clean the burette after use. The final rinse should be with deionized water. Drain a few mL through the tip to avoid clogging.

APPENDIX E

GENERATING A STRAIGHT LINE GRAPH IN MICROSOFT EXCEL

Note: Microsoft Office 365 must be *downloaded* on your computer to add a trendline to the graph.

1. In Excel enter
   a. *x-axis values* in column A
   b. *y-axis values* in column B
2. Highlight the data in the columns. Select the **INSERT** tab at the top of the spreadsheet. In the **Charts** menu, select the drop-down arrow for the **Insert Scatter (X, Y) or Bubble Chart** option. Select the scatter chart type in the <u>uppermost left corner</u>. When the cursor is hovering over the correct chart, the name of the chart is simply **Scatter**, ...
   - <u>Do not select</u> **Scatter with Smooth Lines and Markers,** and
   - <u>Do not select</u> **Scatter with Straight Lines and Markers.**

3. Use the green plus sign (➕) near the upper right corner of the chart to select only the following four **Chart Elements**:
   a. Axes
   b. Axis Titles
   c. Chart Title
   d. Trendline (It should default to Linear; if not select it from the menu that appears when hovering over the word Trendline.)
   *<u>Remove checkmarks from other</u> **Chart Elements**.
4. Add titles by clicking directly on the words **Chart Title** or **Axis Title**; a text box should appear. Type your chosen text. Capitalize important words in your title and include units in parentheses when there is one.
   a. Axes titles are straight-forward, for example, *Volume of NaCl Solution (mL)*
   b. Chart titles should concisely state the purpose of the graph, or what is being determined, for example, *Density of Sunkist Orange Soda.* **Note:** Do not use the experiment title as your graph title. Also do NOT use *Mass vs Volume, which is restating the axes titles.*
   c. Your name should be in the **Chart Title** in a distinctly <u>smaller font</u> on the second line.
5. If the trendline does not fill most of the plot, one or both axes minimums need to be adjusted. To do so, right-click directly on the axis and select **Format Axis...** from the drop-down menu; In the **Format Axis** window under **Axis Options** and **Bounds**, change the **Minimum** to a value close to, but below, the minimum in your data set. For example, if the lowest value in your data set is 36.25, a minimum of 35 for the axis is a good choice. Repeat with the other axis if needed.
6. Right-click directly on the trendline and select **Format Trendline....** In the **Format Trendline** window under **TRENDLINE OPTIONS**, select **Linear** and **Display Equation on chart;** click and drag the equation somewhere in the plot area to make the equation more visible.
7. Right-click in a blank area of the graph such as the upper left corner. Select **Move Chart ....** In the **Move Chart** window, select **New sheet** and enter **Chart1**. Select **OK**.
8. In the **PAGE LAYOUT** tab at the top of the screen, in the **Page Setup** menu, select **Orientation** and change to **Landscape** if it is not already selected.

9. Print the graph from **Chart1**.

An example graph is shown in Figure 1. Remember that your graph should fill the page.

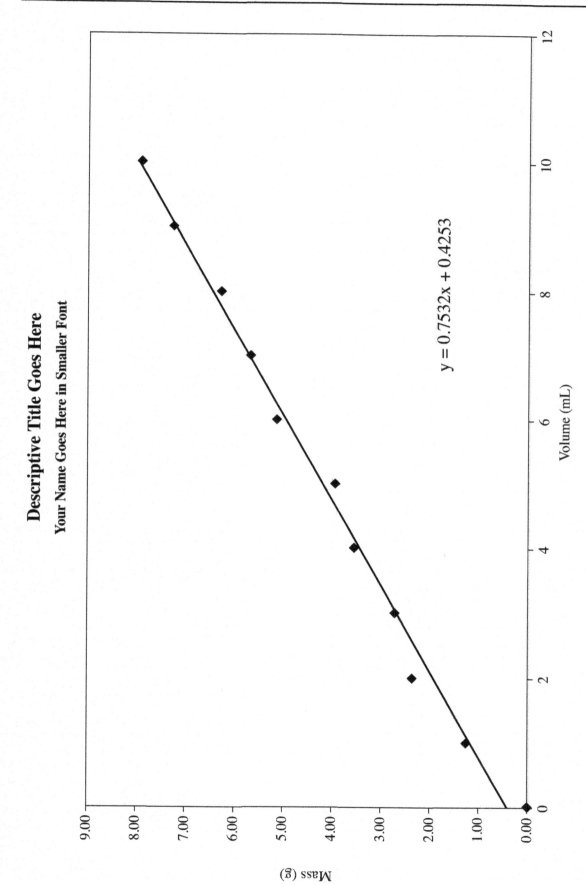

Figure 1 A sample graph